Praise for *Among the Islands* and Tim Flannery

"One of the world's greatest zoologists . . . who's probably discovered more new species than Darwin. He's a remarkable man." —Redmond O'Hanlon

"Flannery is not just an internationally acclaimed zoologist; he's also an adventurer and storyteller who has discovered creatures no other human has seen. His latest record of exploration traces the beginnings of his career during the 1980s and takes him through more than a decade of study in remote islands of the South Pacific. . . . Flannery's writing is generous and exuberant. His enthusiasm is enough to infect even the least science-minded of readers. . . . A breathtaking, informative tour of faraway lands." —*Kirkus Reviews*

"Tim Flannery is a crackerjack storyteller." —*Publishers Weekly*

"A story told with Flannery's signature clarity and lively readability. . . . this is scientific adventure-writing at its finest." —*Open Letters Monthly*

"[Flannery] focuses on the big picture, writes in bold language and broad strokes. . . . The secret . . . seems to be confident knowledge joined to a storyteller's gifts and a writer's determination to get it just right—a rare combination, and a powerful one." —Thomas Hayden, *Washington Post*

"Flannery . . . is without question an extraordinary scientist." —Bill Chameides, *Science Magazine*

"There is an eighteenth- or nineteenth-century explorer inside Tim Flannery trying to get out. And doing a damn fine job of it too in this rather ripping yarn, which combines a passion for science with a zeal for adventure." —*Weekend Australian*

Other books by Tim Flannery

The Weather Makers

Mammals of New Guinea

Tree Kangaroos: A Curious Natural History
with R. Martin, P. Schouten and A. Szalzay

Possums of the World: A Monograph of the
Phalangeroidea with P. Schouten

Mammals of the South West Pacific and Moluccan Islands

Watkin Trench 1788 (ed.)

The Life and Adventures of John Nicol, Mariner (ed.)

Throwim Way Leg

The Birth of Sydney

Terra Australis: Matthew Flinders' Great Adventures in
the Circumnavigation of Australia (ed.)

The Eternal Frontier

The Explorers

A Gap in Nature with P. Schouten

Astonishing Animals with P. Schouten

Chasing Kangaroos

The Future Eaters

Now or Never

Here on Earth

TIM FLANNERY
AMONG THE ISLANDS

GROVE PRESS
New York

Grateful acknowledgment is made for permission to reproduce the following illustrations: Kula canoe, *Sunbird*, outrigger canoes: M. Holics. Kula man: T. Ennis. Woman holding wallaby, Woodlark cuscus, base camp, woman with pipe, Naufe'e, Folofo'u, flowers, Araucaria: Tim Flannery. King rat, Tim with bat: M. McCoy. Poncelet's giant rat: T. Leary. Monkey-faced bat: P. German. Every effort has been made to trace copyright material. Where the attempt has been unsuccessful, the publishers would be pleased to rectify any omission.

Copyright © 2011 by Tim Flannery

First published in Australia in 2011 by The Text Publishing Company

Printed in the United States of America

ISBN 978-0-8021-2182-0
eISBN 978-0-8021-9404-6

Grove Press
an imprint of Grove/Atlantic, Inc.
154 West 14th Street
New York, NY 10011

Distributed by Publishers Group West

www.groveatlantic.com

14 15 16   10 9 8 7 6 5 4 3 2

To ALS

Contents

Bismarck Sea

Bismarck Islands

PAPUA NEW
GUINEA

D'Entrecasteaux Gro

Cairns ●

Coral Sea

AUSTRALIA

0   200   400   600   800   1000

Kilometres

Brisbane ●

Pacific Ocean

THE SOLOMON ISLANDS

Santa Cruz Islands

VANUATU

FIJI

NEW CALEDONIA

Introduction

For over a decade during the 1980s and 90s, I had the best job in the world. I was leading a team of researchers, and together we travelled the tropical islands of the southwest Pacific looking for marsupials, bats and rats found nowhere else on Earth. It was pioneering work. No summary of the mammal fauna of the region existed, so we had nothing to guide us apart from brief accounts scattered in the scientific literature. Many of the islands were a great blank, not having been visited by anyone interested in mammals since some pioneering naturalist dropped by during the age of sail.

Island biodiversity is exquisitely vulnerable to human disruption. The fate of New Zealand's biodiversity is typical. Over a third of its land bird and bat species have become extinct since human settlement, and another third are threatened with extinction. Such figures made me wonder how the mammalian inhabitants of islands further north were faring in the face of introduced species and European colonisation. Because nobody in recent times had gone to the islands to look, it was a question without an answer, and it became the *raison d'être* for our quest. It was possible, we knew, that some species had vanished before anyone even realised that they were endangered, making our adventures somewhat quixotic. But it also seemed possible that a species or two existed among the network of islands, reefs and misty peaks that had evaded earlier visitors, and so awaited scientific discovery.

Planning for our expeditions was carried out in musty libraries and museums, for these were the days before the internet, and if one wished to consult the pages of the journal of the *Museo Civico di Storia Naturale di Genova*, or the obscure, Shanghai-published journal *Memoires concernant l'histoire naturelle de l'Empire Chinois*, one had to front up at a library that held the volumes—and often-times ask for help from a translator as well. But many important finds were not even mentioned in print. Perhaps their discoverer had died among the islands and his trove of treasures had been shipped home to find nobody knew enough, or cared enough, to publish an account of what was in them. And so we worked among often forgotten collections held in museums great and small, from London to Beijing, poring over stuffed skins of rats and bats for evidence. Many had somehow survived fire, war and penury, and as we chased down century-old clues to the existence of such bizarre creatures as giant rats, monkey-faced bats and piebald cuscuses, we stood in awe of those who had built the collections, as well as the curators who had tended them through decades or centuries of peril.

To open the drawer of an old museum cabinet and find the remains of a creature that lived on a distant tropic isle long since dramatically changed by European impact, and which had travelled halfway round the globe to reach its keeping place, is a magical experience—like travelling in time itself. That moth-eaten skin, or even fragment, may be all that remains of an entire species, tinting the thrill of seeing it with immense sadness. For here lies all, perhaps, that is knowable of a branch of life that may have gone its own way for a million years or more, a life form that once played an important role in an island ecosystem, but which had now winked out, never to be seen again.

Or had it? Who could really say whether it might still survive in the densest jungle or remotest peak of its island realm. The most elusive specimens lacked even a clear indication of where they'd been collected: their labels cited a whole island group, or even the voyage of discovery on which they'd been collected, rather than a precise island as their origin. Such species present a great challenge, but even the better known kinds presented us with a daunting quest. Where, and how, does one begin on an island the size of a European state to search for a fist-sized, nocturnal creature that has not been sighted for over a century? But we were young, and confident that we could make sense of the clues. Sometimes guided by no more than a single word on a specimen label, we found our way to unimagined places—to villages that had not seen a white face in living memory, or mountaintops crowned with surreal vegetation—and then we knew that the quest was at least as important as the goal.

As our work progressed, the high peaks of the Pacific Islands began to hold me in their thrall. From a biological perspective they are among the least known places on Earth, and even today some island peaks—which rival Australia's Mount Kosciusko in elevation—remain unvisited by Europeans. Often held sacred by the native people, these mist-wreathed summits are in some respects lost worlds—islands in the sky crowning islands in a tropical sea.

But reaching them was not easy. Local taboos, foul weather, dense jungle and sheer remoteness combine to make them some of the hardest places on earth in which to conduct biological research.

The great arc of islands between Sulawesi and Fiji was our stomping ground. Extending over 6000 kilometres, and crossing the Equator, it's a huge realm: stretching further than the distance from Paris to Montreal, and almost as far as Beijing to Cairo. But it's different from any similar-sized region in that it consists of thousands of islands, each distinctive in its geology, vegetation, shape, size and history of human colonisation. From the picture-postcard Polynesian atoll, to some of the largest, highest, most rugged and ancient islands on Earth, the region truly is a version of the world writ small.

Among these varied lands are some that originated as slivers wrenched from ancient supercontinents a hundred million years ago. Others are continental chips torn more recently from the great island of New Guinea. Yet others formed as volcanoes that belched forth from the ocean depths, arriving above the waves as virginal lands innocent of life until drifting seeds, spores and insects arrived. Krakatoa, which blew itself apart in a paroxysm of volcanic activity in 1883 and then grew anew from the sea, gives us some idea of the process. First ferns and insects, then flowering plants, birds and lizards arrived to colonise their new-found land. Krakatoa is just a few tens of kilometres from Java and Sumatra, and less than a century old. Imagine a volcano surfacing a thousand kilometres from the nearest land, then receiving its pilgrims over a million years.

There are other ways, too, for islands to form. Some are simply heaved above the waves by the movements of continental plates, while others, known as land-bridge islands, were, just 10,000 to

20,000 years ago, attached to much larger landmasses by connections now drowned in a rising sea. Some islands originated as combinations of several of these processes. But all islands, regardless of their origins, have one thing in common: transience. True, some islands are longer-lived than others, but all, in the unfolding of geological time, are destined to sink beneath the waves or amalgamate with larger landmasses. Over just the past few centuries, dozens of islands have been born and have died, and like us ultimately all of them will die, even as new islands are born.

On islands, evolution can be slowed down or speeded up. It can also take unlikely directions, fashioning novel creatures adapted to the particular conditions of the island. Why do islands have such peculiar powers over the evolutionary process? Imagine taking a species from the complex, rich continental ecosystem in which it has evolved, then releasing one or two randomly chosen individuals on an island where nothing like it has previously existed. If they survive, the individuals that begat the island population will have just a small subset of the species' genetic diversity, and this alone will wield an influence. To understand how, just imagine choosing two humans—say a red-head and a very tall person—and leaving them on a desert isle, then returning in a million years to examine the characteristics of their descendants.

But genetics is only the start, for when any creature reaches an island, it has effectively been transplanted into a new world. Its predators, competitors, diseases and even favoured foods may not exist in its new home, and, instead of the boundless habitat of a continent, it finds itself part of a tiny population hemmed in on each side by sea. Such circumstances can greatly speed the evolutionary process. In the beginning the species is likely to breed up rapidly, for in the absence of predators and disease there is no check on its rate of increase. But soon it faces overpopulation: most individuals will die and only those with a particular advantage will survive. Perhaps those lucky few can utilise some food inaccessible to the rest, or perhaps they can conserve energy because they don't fly much, or perhaps they are smaller than average and can subsist on the slender resources the island affords them. Because the population is small and the selection of the survivors so rigorous, the evolutionary process is greatly accelerated. The influence of such powerful evolutionary pressure can be profound. As we see in the dodo, sometimes it creates beings which don't appear to belong here on Earth. Hardly anybody studies mammals on islands: it's those able colonisers the birds that usually get the attention. Yet some island mammals have, like the dodo, been as remarkably transformed. On islands, bats have been known to take on some of the characteristics of apes, and rats those of badgers, shrews or possums. Even humans and their behaviours are shaped by island life, and as a result island cultures have become as varied and novel as any on Earth.

Not all island life is established by vagrants arriving by raft, wind or wing. Islands that originate as slices of continents, severed from the mainland by powerful geological forces, carry with them a subset of continental life. When that happens, entire ecosystems are set adrift to adjust over millions of years to life in a small, isolated sphere. Inevitably, some species become extinct, unable to adjust to the limited circumstances they find themselves in. At the same time new species, which evolved in other parts of the continent, are unable to invade the now-isolated island. The result, for the island survivors, is often a slowing of evolutionary change. Because competition drives evolution, fewer competing species means less change, and in these circumstances evolution can come to a near halt. So it is that islands can become arks full of 'living fossils'—species whose relatives elsewhere are long extinct or transformed by evolution into very different kinds of creatures.

Evolution on islands also plays with the size of creatures. The world's islands are (or rather were) full of giant rats and tortoises, and oversized flightless birds. But there are island dwarves too. Before humans arrived, elephants, mammoths and hippos the size of Shetland ponies abounded on some islands in the Mediterranean and Arctic. There was even a miniature hominid, the hobbit, on the island of Flores in the Indonesian archipelago. Such gigantism among the tiny, and dwarfing of the great, means islands are great levellers, the species isolated on them converging on an ideal size.

Whatever their size or origins, the species and the ecosystems of islands are exquisitely sensitive to invasion. Whether arriving by themselves or by human agency, invaders can be fatal to the original inhabitants, and virtually no island survived into the last century unscathed. After humans the most common invaders are introduced rats—though cats, snakes and even snails have time and again devastated entire islands of unique species.

Why should the long-isolated islanders be so vulnerable? Island species often have only a limited range of predators, competitors and diseases. Under such circumstances, any species that invests too much energy in being able to flee swiftly or in producing toxins that deter predators will, prior to the arrival of new invaders, be outcompeted by those that put less effort into these faculties and more into reproduction. This is why so many island birds are flightless and why the nuts and leaves of so many island trees are edible, even without cooking—trees that invest in toxins where there are no predators, rather than making more nuts and fruit, are disadvantaged in the evolutionary race for survival. But island species also lose their fear. Fearful creatures use a great deal of energy in fleeing danger, both imagined and real. Where the danger is almost entirely imagined, evolution selects for individuals that conserve their energy for reproduction. Island birds have been known to sit on their nests even while being eaten alive by rats. Many will not flee from cat or human, even when attacked.

As colonial history has shown, the native cultures of islands are also vulnerable to change from outside. The power structures of Hawaii were transformed by a dozen or so iron blades fashioned by a ship's cooper during Captain Cook's voyage of discovery. So armed, the few fortunate chiefs who received them went on to forge empires. What is remarkable, however, is how much traditional island culture still survives in spite of the constant waves of challenge arriving from the larger world. Change may be integral to island life—as old as the first island and as pervasive as the sea that surrounds them all—but such examples convince me that, given half a chance, much that is unique to the islands can persist.

The continent of Australia has cast a long shadow over the extensive archipelago we explored—a shadow composed of living things that, over the aeons, have drifted, flown or made their way some other way to the islands. Among them are curious marsupials, unique birds of paradise and countless other life-forms which have their points of origin millions of years ago on the wide brown land. But what a sea change was wrought on their descendants as they settled into island living. Such species are utterly irresistible to biologists because their myriad modifications reveal the secret workings of evolution. When the animals' ancestors left, Australia's rainforests were extensive, and a vestige of ancestral forms long vanished from the larger landmasses can be found on some islands. Examining such plants and animals can reveal living clues to a now vanished and very different Australia.

We twentieth-century biologists travelled through the vast island realm that was our field of research by whatever means were at hand—sometimes by air, at others by ocean liner, inter-island ferry or dugout canoe. Then we would set about collecting, documenting and exploring a world of nature that in some cases had never before been entered by a biologist. And the excitement of setting up a mist-net and laying a trap-line on an island which had not yielded a single record of a mammal was just about the most exciting thing you could do.

I

IN THE REALM OF THE *SUNBIRD*

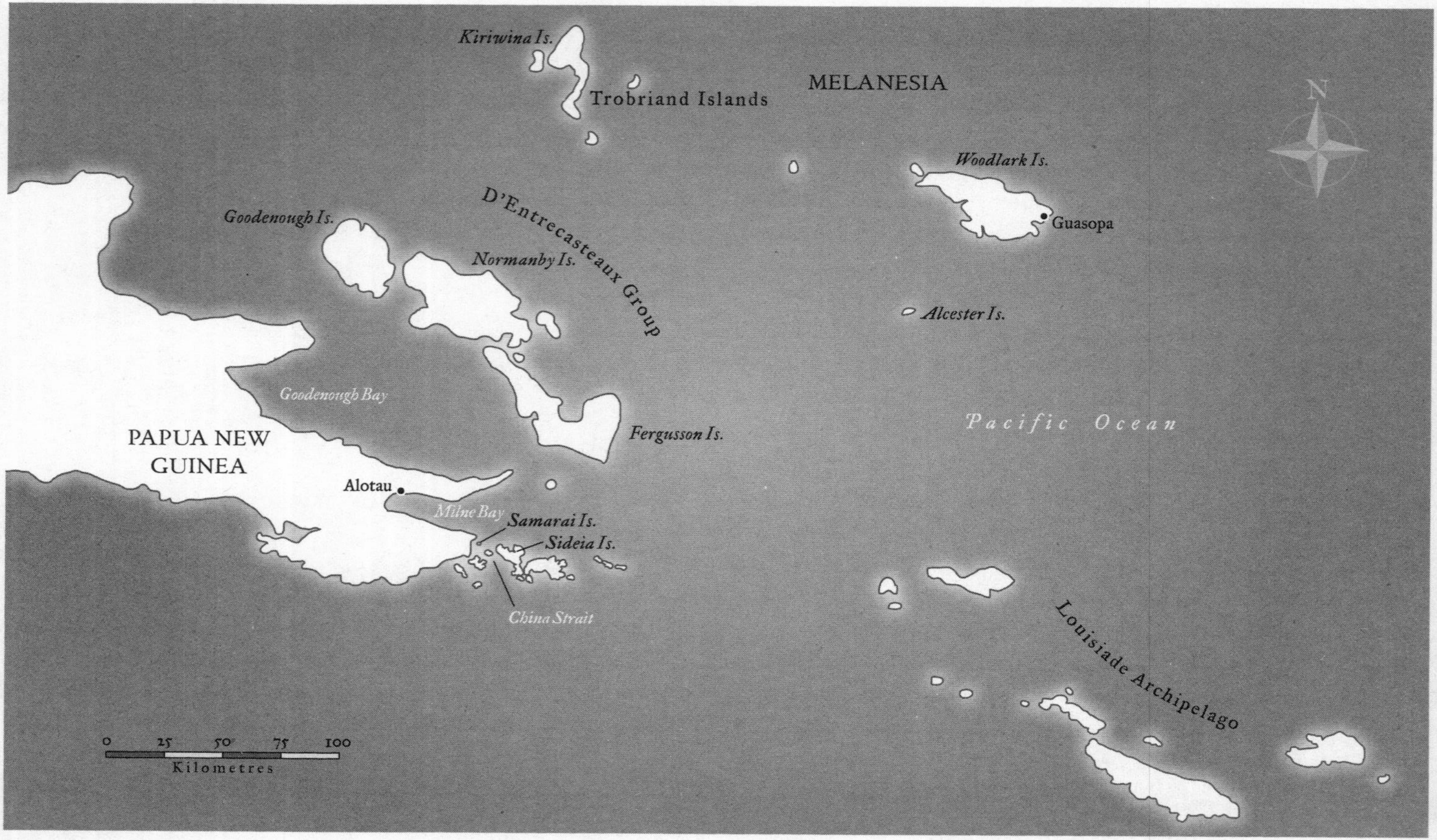

Kiriwina Is.
Trobriand Islands
MELANESIA
N
Woodlark Is.
Guasopa
D'Entrecasteaux Group
Goodenough Is.
Normanby Is.
Alcester Is.
Goodenough Bay
Pacific Ocean
PAPUA NEW GUINEA
Fergusson Is.
Alotau
Milne Bay
Samarai Is.
Sideia Is.
China Strait
Louisiade Archipelago
0 25 50 75 100
Kilometres

The diverse islands that lie off southeastern New Guinea are linked by a common human culture. This is the region of the Kula Ring, where men have been sailing magnificent canoes around the islands in order to trade shell valuables known as Kula since time immemorial. It's a fascinating region whose biodiversity, while limited, is still being discovered. And it was there that I first came among the islands.

It's odd to think that parts of Melanesia had been mapped by European explorers before the interior geography of their own countries had been accurately portrayed on paper. But, if you sail, the ocean is a highway. Using it, the ancestors of the Polynesians colonised two thirds of the southern hemisphere—from Madagascar to Henderson Island in the far eastern Pacific—leaving behind them civilisations with common linguistic and cultural origins. And a millennium or so later the Europeans followed them, first mapping and charting archipelagos, then settling outposts and colonies, the islands of southeast Papua being among the last to be colonised.

The largest and biologically most important islands of southeastern New Guinea comprise the D'Entrecasteaux Group. They lie just off the northern coast of New Guinea's eastern end, forming a string of large islands 160 kilometres long. The first European to sight them was the Frenchman Antoine Raymond Joseph de Bruni d'Entrecasteaux, captain of *L'Esperance*, who sailed in search of the lost Lapérouse Expedition. It had not been heard from since it sailed out of Botany Bay in early 1789, and its fate—shipwrecked off the island of Vanikoro—would not be revealed for decades.

Despite the fact that the island group bears his name, d'Entrecasteaux did not step ashore, and so he recorded very little. It was Captain Moresby who in 1874 first made European landfall and soon thereafter missionaries, merchants and biologists followed.

Lying directly east of the New Guinean mainland is a string of islands known as the Louisiade Archipelago, which was named in 1768 by another French Mariner, Louis Antoine de Bougainville. The importance of French exploration and mapping of the Pacific is not widely appreciated today, though a look at any map of the area will reveal a significant legacy of French names spread liberally across the region. The achievements of the French explorers rivalled those of the British, yet, curiously, their great colonial possessions of Tahiti and New Caledonia were not discovered by French explorers, but by the English.

During our expeditions we only touched upon the western end of the Louisiade Archipelago and to this day its mammals remain poorly documented. Between this extensive chain of islands and the mainland lies China Strait, and in it Samarai Island. China Strait is a major shipping route. It is well known to mariners, and so, once, was Samarai Island, which sits in the shipping lane. At just twenty-four hectares, it's a mere speck, yet by 1907 it was the regional seat of government. Then it supported three pubs, had its own bishop, rectory and church, three trade stores, as well as various government buildings, hospitals and private residences. By 1927 electric power and street lighting had been installed and Samarai Island was on its way to becoming a thriving regional centre. But it was destroyed, in part by its own inhabitants, in January 1942. Fearing imminent invasion by the Japanese, they tore down or rendered inoperable all useable infrastructure.

Beyond these archipelagos lie the more isolated islands, which include Kiriwina in the Trobriand Islands, Woodlark and Alcester.

CHAPTER 1

Woodlark, the Wandering Isle

It was the lure of being able to travel in time as well as across the seas that first carried me to the islands. It was 1987 and I was in my early thirties, and I must admit that daydreams of Polish anthropologist Bronislaw Malinowski's fabled Islands of Love were part of the attraction. Malinowski had lived on Kiriwina Island in the Trobriands Group during the 1920s, and in *The Sexual Life of Savages* he'd reported in lively terms on the seemingly promiscuous young people he'd found there.[1] At the time I was in charge of the mammal department at the Australian Museum in Sydney, and while girls in grass skirts are, technically speaking, mammals, they most emphatically did not fall within my research remit. Instead it was the distributions of possums, bats and rats that would need to set the agenda.

I'd been employed as a research officer. In pay and status it was the bottom of the scientific totem pole, but I couldn't have cared less. What mattered was that I was expected to carry on the great tradition of the curators who'd preceded me and conduct research on the mammals of New Guinea and the southwest Pacific Islands. The museum had a proud history in the Pacific, and I soon learned that its collections held many important specimens, some acquired during the age of sail. As I pored over the mammal collections, the rudiments of a geography of the distribution of Pacific Island mammals began to take shape in my mind. But even when combined with what I learned from published sources, the picture was woefully incomplete—like a jigsaw with nine out of every ten pieces missing. A new book on the mammals of Australia had just been published, and it occurred to me that there should be a similar book for the southwest Pacific. But with so many ghastly blanks in our knowledge, an immense amount of fieldwork would be required before I could put pen to paper.

While full of goodwill for my aspirations, the museum could provide no financial support for my plans beyond my modest wage. So it was clear that I'd need to find a source of funding. And this, in the case of the Islands of Love, as they became known, was provided by TAMS—the Australian Museum Society. Headed by the wonderful and ever-gracious Susan Bridie, the society consisted of several hundred mostly well-heeled supporters of the museum, some of whom who were keen to participate in scientific research.

And so it was that on a bright, breezy August day in 1987 I found myself standing on a wharf in far north Queensland alongside a group of people I hardly knew. It was the season of the southeast trade winds, the sea was covered in whitecaps and the wind was relentless and salty. A mountainous pile of scientific equipment—from traps and nets to supplies, along with great silver vacuum flasks of liquid nitrogen (so that we could store samples of DNA that might reveal how the creatures we

hoped to encounter had reached their island homes)—lay on the
dock beside an aluminium catamaran with the name *Sunbird* on
her bows. She'd been purchased for the Australian Museum by
Suntory, the Japanese whisky producer, though just why alcoholic
beneficence had smiled upon the museum in this particular guise I
never discovered. Perhaps some past director was a great supporter
of their product, I mused as we heaved the gear aboard.

The whole business of the expedition seems ridiculously romantic from this distance in time. TAMS had seen fit to organise and
fund the work, the *quid pro quo* being that five of its members
would participate in this voyage of biological discovery. Our
objective was to survey one of the most inaccessible large islands
in Melanesia—Woodlark in the Trobriands Group. Woodlark
attracted me because of its size, its small human population and
the abundance of undisturbed habitats. Furthermore, it was home
to an unusual cuscus (a cat-sized marsupial), and I felt that other
biological novelties might lie hidden there. But it lacked an air
service—hence the need for the *Sunbird*.

Woodlark had been visited by scientists interested in mammals
only twice prior to our expedition. In 1894 Albert Meek, one of Lord
Walter Rothschild's most adventurous biological collectors, had tried
to reach the island on a seven-metre whaleboat. He wrote later of
his foolhardy attempt that, 'I had no knowledge at all of navigation
and had not even a compass aboard...I was to learn in the school
of experience that navigation was not a matter that could be taken
in the casual way.'[2] Day after day the understating Meek found
himself driven back by the same trade winds that we experienced
on the dock in Cairns, until he was finally blown far out to sea where
he had to proceed by the light of the moon towards an unknown
shore. Without matches, food or shelter, Meek was forced to

abandon his first attempt to reach the island.

Months later he tried again, but a great wave washed him and his crew clean out of their vessel. He said it was only the pain of being severely cut about the legs by coral that gave him the impetus to save himself from being battered to death. But an Aboriginal boy who was travelling with him perished on that jagged shore. Clearly, a more substantial vessel was required, and so Meek purchased a nine-tonne cutter. Upon finally reaching Woodlark in 1895 he found what was, from a zoological point of view, a virgin island: nobody had previously collected there. When I first read Meek's classic account of his experiences, *A Naturalist in Cannibal Land*, I was hoping for a rich word-trove of experiences and observations, and was appalled to discover that he dealt with the island in just four lines.[3] Perhaps he was too exhausted or ill to write more, or perhaps he found the place dull. Whatever the case, among the discoveries he made on this most inaccessible isle (which he fails to mention in his book) was a peculiar kind of possum whose coat was spotted in black, brown, yellow and white like a crazy patch-work quilt. Remarkably, every individual had a unique pattern—a characteristic seen more often in domesticated rather than wild animals.

Nearly sixty years were to pass before another biologist followed in Meek's footsteps. This time the researchers came from the American Museum of Natural History and were part of a well-organised and well-funded expedition. In 1956 it spent three weeks exploring the south and west of Woodlark Island and reported that the island's lush forests were deficient in mammal species. The researchers thought that the unusual cuscus was very rare. Although they added some bat and rat species to the limited list of mammals collected by Meek, I was far from convinced that they had exhausted all avenues of research. Now, I thought, as I sat on the dock in front of the *Sunbird*, it was my challenge. What, if anything, would our team discover?

As I pondered our prospects, a white-haired, denim-clad figure emerged from the *Sunbird*'s cabin. A pair of slightly sun-damaged but still bright blue eyes shone deep in his weather-beaten face. 'I'm Matt Jumelett, skipper of the *Sunbird*,' he announced in a thick Dutch accent. Then a considerably younger blonde woman emerged from the same hatch. 'And I'm Mipi, the crew,' she added as she extended her hand in welcome. 'Come aboard and have a cup of tea!'

So we expedition members climbed aboard—leaving the loading of the cargo for the moment—to introduce ourselves to the *Sunbird*'s captain and crew. Our team consisted of Aziz Irani, an ever-smiling businessman of Persian descent; Robert Saunders, a publisher seeking adventure; Tish Ennis, a nurse; Des Beechey, a computer expert and amateur shell collector; and Michael Holics, an environmentalist and state-government employee. On Woodlark we would rendezvous with two other expeditioners: biologists Dr Greg Mengden, a Texan and world authority on venomous snakes, and Lester Seri of the Papua New Guinea Department of Environment and Conservation. Lester and I had already undertaken three expeditions together to remote parts of the New Guinean mainland, and we were to become lifelong friends.

Following a cuppa we loaded our gear into the *Sunbird*. It was a tight fit, with boxes of traps and nitrogen flasks in every nook. Matt said that crossing the Coral Sea from Cairns to Samarai would take four days, and that due to the southeast trade winds, which were against us, he expected to make the trip under power. I was disappointed that we wouldn't be sailing, but as it was the first time that the *Sunbird* had left Australian waters and the wind promised a rough crossing, I understood the decision.

Soon we were in it, the wind in our teeth, the hum of the engine filling our ears, and a relentless buffeting of waves against the *Sunbird*'s hull. I slept in the for'ard cabin that night, where the low hum of the motors and the thump, thump, thump of the waves lulled me to sleep. But some time in the wee hours I woke to a thump that sent the whole vessel into a great spasm. In my groggy state I had the sensation that we were going down, down and down—and were not coming up. We had left the protection of the Great Barrier Reef and entered the Coral Sea.

It must have been around 2 am when, too excited to sleep, I made my way onto the deck to be greeted by a sky the likes of which I had never seen before. It was alive with stars spread from horizon to horizon. Phosphoresence trailed in our wake, and every now and again a flying fish would hit the aluminium decking with a thud. It was the kind of night that is more glorious than any day, when bed seems a dull thing indeed. Not everyone was as fortunate, however. Des Beechey was a martyr to seasickness. I was grieved to learn that he couldn't leave his bunk, and he looked more dead than alive when I went to him the following morning. Tish administered a sedative that seemed to relieve his agony, but despite this Des barely left his bunk for the entire voyage. None of us was surprised when he chose to fly back to Australia from Port Moresby at the end of the expedition, rather than endure another sea crossing.

To my landlubber's eyes, our days and nights on the Coral Sea were filled with wonder. We passed mysterious foamy lines in the ocean where currents met, and there we would sometimes see pilot whales and strange sharks floating on the surface. In such places a great mahi-mahi would occasionally take the lure trailing behind the catamaran, and we would pull the sparkling fish aboard. These wedge-shaped predators can be a metre long, and are built for speed, their prey being flying fish. The creatures flapped on the deck and pulsed with life, their iridescent yellow and blue-green colours dazzling us. Empathy is unwelcome when you're fishing for dinner, but how sad it was to see their eyes still and the vibrancy of their colours vanish.

At other times the lure would be taken by a tuna so enormous that it took our combined efforts to drag it in. The mahi-mahi, which are among the most delicious of fish, we filleted and ate, but the tuna were consigned to the freezer. I already knew something of the customs of the people we were to visit: a tuna is a fine gift for a boatful of visitors to bring to an island.

As the voyage progressed I grew close to our skipper Matt. He told me that during World War II he had been in the merchant navy, then he had become a submariner. After many experiences on and below the sea, Europe seemed too small for him, and he had come to the South Pacific where he had joined the renowned trading company Burns Philp. While with them he'd captained every type of vessel in the great company's fleet, from the meanest, most cockroach-infested intra-island copra-boat to their top-of-the-line freighters. What he did not know about New Guinea's waters was not worth knowing, and he exuded an aura of salty confidence that inspired all who travelled with him. But one thing about Matt was clear: after years sailing as the only white man in Papuan crews, he was The Captain. To crew and us alike, his word was law.

I never once saw Matt refer to the *Sunbird*'s navigation system. Instead, he preferred to chart our course with compass and dividers on a map he spread across a table on the bridge. And the man never

seemed to sleep. I tried to join him on the night watch, and would
listen as he told the most outrageous stories—of sinking smuggled
hand-guns into barrels of grease to sell to Chinese merchants, of
tidal bores on the Fly River that could sink a ship, and of near
misses with reefs and hurricanes in vessels that should not have
been afloat—but I could never last the night. And, when I awoke,
Matt was always still there, sipping on a coffee as his eyes scanned
the northern horizon in solemn concentration.

Our fourth evening aboard the *Sunbird* was balmy, and for the
first time the wind had died away. Clouds had drifted in. Matt
quipped that the night was as dark as the devil's arsehole. As I sat
propped against the mast and blanketed in the black velvet night,
my nose was assaulted with an unmistakable, acrid smell—the
smoky odour of a New Guinea house fire. It was six years since
I'd first visited New Guinea, but I knew the smell well. It gave me
the sensation that I'd been transported instantly to land, and was
crouching before the fire, surrounded by black skin and flashing
eyes. An hour or so later the odour of the smoke was joined by
another immediately identifiable scent: tropical vegetation decay-
ing in the eternal dank of a sago swamp. After four days at sea my
nose was sharper than it had ever been, conveying the proximity
of both village and bush on the mysterious island of New Guinea.

It was long after dark when we reached our anchorage in China
Strait, at the far eastern end of New Guinea. There we awaited the
coming of the dawn and a customs officer. For some time I'd had
qualms about meeting this official. Quite apart from the mountain
of scientific equipment I'd have to explain, we'd purchased duty-
free alcohol rather enthusiastically in Cairns, and a sizeable excise
bill seemed likely. Matt, however, remained imperturbable, and at
the approach of the customs launch at around 8 am he greeted the

officer fluently in pidgin English as if he were a long lost brother.

As soon as the crisply uniformed fellow was aboard, Matt offered him, by way of breakfast I assumed, a cold can of Foster's lager. This was gratefully accepted and drunk, at which point a second was immediately offered. Shortly after, Matt opened the locker holding our extensive supply of duty-free alcohol, and to my great surprise nothing was said about duties payable. As to our equipment, it was evidently nothing out of the ordinary. Either that or Matt had skilfully diverted the officer's attention with another beer.

Thus far things had gone smoothly, but then the customs officer requested our passports. The research team had theirs at the ready, but Matt fiddled about and said, again in fluent pidgin, 'Passports? What's this about passports? That's new to me!' It was twelve years since Papua New Guinea had declared its independence from Australia. I was horrified, expecting that we would be sent packing and the entire expedition aborted. How, I wondered, would I explain this to the museum director? But, to my astonishment, a saintly sort of long-suffering look began to take shape on the customs officer's face. Then he sighed and murmured half to himself, 'Taim bilong masta.' In pidgin, the phrase refers to the colonial era, and conveys both fond memories of a period when government worked, at least vaguely, and irritation at the often high-handed ways of the colonial masters. Perhaps our customs officer had had a good colonial experience, for he turned to our passport-free captain and said in English, 'Next time you visit our country, you really should bring your passport. We're independent now!'

After this seemingly miraculous customs clearance, we went ashore at the tiny island of Samarai to buy food and to stretch our legs. The town is so small that you can walk around it in minutes. Yet it was once of considerable economic importance, for it is located in the China Strait—the keyhole through which most trade between China and Australia passed during the age of sail. And there, awaiting the passing ships, were the accumulated riches of Papua. Copra, pearls, trochus shell and bird-of-paradise plumes once filled its ample storehouses and cluttered its docks. But by the time of our visit the emporiums of Samarai lay locked and rusting, all but empty. Slowly tropical nature was reclaiming the island as its own. Saplings were sprouting between the sheds, and the submarine pylons of the dock were clad in languorous sea-fans and elegant long-spined sea urchins, around which flitted clouds of tropical fish so bright and sprightly as to take one's breath away.

Not all trade is gone from Samarai, however, for in place of the pith-helmeted traders of yesteryear we found shy Papuan women. Some sat before the derelict stores, their wares meticulously laid out before them. A piece of brightly patterned cloth might display a fan of betel nuts, lime, *daka* and piper leaves—all the necessaries for that great Melanesian pastime of *buai* chewing—laid out with a precision and elegance that would do a great department store proud. Others held just a dozen or so small limes, while others displayed tropical seashells. At the news of these last offerings, Des Beechey rose from his bunk of misery and purchased a few of the more unusual ones. For a moment I felt that the trip might have been worthwhile for him, but Des's parole was brief. With our legs stretched and our curiosity satisfied, the *Sunbird* set out northward, across the white-capped Solomon Sea, forcing Des once more into his bunk. He was, I thought, enormously brave, for he could have disembarked at Samarai and made his way home.

To approach a tropic isle by sea is a matter of great anticipation.

At first, perhaps only a cloudbank might be visible—a chimerical sign that could evaporate into nothing. But below it might indeed lie the desired island. A slender grey smudge hugging the horizon might next be glimpsed—still an uncertain sign which might transform into a reef or current. But if the smudge takes shape, and the peaks, forests, reefs and bright beaches hove into view, the journey's end is in sight. Thus was our approach to Woodlark. Muyuw, as its people know it, is large, mysterious and isolated, with most of its 800 square kilometres covered in primary rainforest. Arriving felt like drifting a century back in time.

What I did not realise then is that Woodlark had been travelling too, and along roughly the same track as the *Sunbird*. As anthropologist Fred Damon says of the island: 'I think it needs to be stressed that Woodlark Island is in motion.'[4] Its geological origins lie east of Samarai, among the islands trailing off southeast New Guinea. Our journey had taken just thirty-six hours, but Woodlark's had taken millions of years. As it drifts serenely northward on its submarine rise, its mountains are slowly rising in parts and falling in others. Such movements cause earthquakes, and for an entire month in 1914 the place suffered such violent tremors that the natives slept by their canoes, fearing that their island home would topple into the sea.

In the south, Woodlark's forty-million-year-old volcanic rocks are folded into rugged hills up to 400 metres high, and rainfall is high. In the east, where we landed and where most of the people live, Woodlark consists of raised limestone and a distinct dry season prevails. The prehistory of the island remains mysterious. Three great stone ruins—the island's own Stonehenge, formed of great slabs of carved volcanic rock—lie at its centre, but nobody knows how or why they were made.

Woodlark was one of the last of the world's larger islands to be charted by Europeans. It remained unsighted, or at any event uncharted, until around 1832, when an otherwise forgotten Captain Grimes of the Sydney-based whaler *Woodlark* noted its existence in his log. Gold was discovered in 1895 (the island still has a small goldmine) but by far the biggest intrusion from the outside world occurred in June 1943 when the American 112th Cavalry arrived and built an airstrip and barracks. With them came tonnes of material then utterly unknown and unimaginable to the locals. Despite these disruptions the people of Woodlark continue to participate in the famed Kula Ring, which involves men making long journeys by canoe to other island groups in order to exchange shell valuables.

We were still at sea. Between us and our anchorage in front of Guasopa village lay a narrow, winding channel dotted with potentially fatal coral bommies. The sun was low on the horizon, and even with sunglasses on it was hard to see through the glare, so I climbed the mast and scanned the channel leading to the beach ahead. All around, the water was a thousand shades of blue— some as vivid as I had ever seen—and beyond it lay rugged, green Woodlark, its peaks obscured in cloud, its forest cover uninterrupted as far as the eye could see. What creatures lay hidden there? What mysteries awaited us? Late in the day as it was, I was determined to go ashore the second the anchor hit bottom.

We finally reached water above a patch of clear, white sand immediately in front of the beach at Guasopa. It was protected by a point from the prevailing wind, and was so unriffled and clear that it seemed we were floating on air. Every detail of the bottom four metres below was visible. Delighted, Matt growled at the crew in his best gruff, Dutch-accented, captainly way to, 'Put down the anchhorr!' But instead of the sound of rushing feet hastening to obey his command, nothing was heard but a thin, dreamy female voice saying, 'But Matt, doesn't it look so much nicer over there?'

A prodigious thunder cloud crossed our captain's visage. Never in all his years of sailing had an order been so impertinently countermanded! I watched as he struggled to regain composure, to remember that his crew was now his attractive young wife. And then I did something mischievous. By now the *Sunbird* was surrounded with outrigger canoes manned mostly by children curious to know who was visiting their village, and in an ill-advised moment of levity I quipped, 'Well at least the locals know who the *Sunbird*'s real captain is.'

The explosive nature of the eruption that followed was of a kind not seen before on that volcano-free island. The torrent of Dutch was so fluid and voluble that it brought to mind the *nuée ardente* that consumed Pompeii, and it scattered the outriggers floating around the *Sunbird* to the four winds. Finally Matt roared like a wounded bull, 'Mipi, now they know who the real captain is!' In the silence that followed the anchor was released, and hit the sand below with remarkable alacrity.

We hitched a lift to shore with one of the outriggers, and within minutes felt the gritty sand of Woodlark Island beneath our feet. The village was almost on the beach, and I quickly found the head of the village council and explained our business to him. He greeted us enthusiastically and granted permission for us to work. He also mentioned that there was already a group of researchers in the village. Over the last century only two biological expeditions had visited the remote island. What were the chances that the third and fourth should arrive simultaneously, both intent on studying the island's cuscus?

While the astonishing news dented our explorer's pride a little, the presence of other researchers turned out to be a great benefit to us, and led to the development of lifelong friendships. Chris Norris and his team of students from Oxford University had planned and paid for their own expedition. They were particularly interested in the ecology of the cuscus, while we were more interested in its evolutionary relationships.

Because much of our work took place at night, and because the *Sunbird* was soon to leave in order to pick up other expedition members, we needed a base on the island. The head of the village council had allocated a disused shop for our accommodation and makeshift laboratory. Its concrete-slab floor was not comfortable, and it was full of mosquitoes, but at least it was secure. In light of the dangers that our liquid nitrogen fridge and formaldehyde presented to curious children, this was an important consideration.

As we explored the area we discovered that the village of Guasopa is built over the remnants of a vast American army base. The base was used only briefly after being established in July 1943, because by the end of that year the focus of the war had shifted westward. Yet the locals remembered it clearly, especially that there had been both black and white troops, whom they recalled being buried in separate cemeteries. The jungle had reclaimed most of the old infrastructure, but parts of the airstrip surface—which had been made of living coral—remained solid, and many of the villagers had built their houses on it.

I was surprised to see how dry the ground was around Guasopa. The thin forest canopy allowed the sun to penetrate, and the limestone immediately drained away any rain that fell. The trees had dropped many of their leaves, and we walked everywhere on a crackling carpet of dry leaves and twigs, which is hardly ideal for spotlighting flighty animals like possums and bats. Later we learned that we'd arrived in a ferocious el Niño-fuelled drought that held all of eastern New Guinea in its grip.

Our primary objective was to make a thorough survey of the mammals of the island, and that meant setting cage traps for rats, spotlighting for cuscuses and other creatures, and visiting as many caves as possible to look for bats. Once we had unloaded our equipment and set about this work, we sent the *Sunbird* to Kiriwina Island to pick up snake expert Greg Mengden and biologist Lester Seri, who had flown to the airstrip there and were coming to assist us. Back at the Australian Museum in Sydney, I had promised Greg to hold whatever snakes the islanders brought us until his arrival. Greg stressed that the creatures had to be alive—he was anxious to take samples for molecular analysis.

When I explained this part of our mission to the head of the village council, his eyes grew wide, for the islanders generally fear and avoid snakes. But word of the impending visit by this remarkable person and his need for living serpents, preferably venomous ones, spread as quickly as news of a circus coming to town. That very afternoon, snakes—tied, trussed and bagged in every way imaginable—began pouring in, and soon our humble abode was festooned with dozens of canvas bags holding writhing and understandably angry serpents. Indeed, I suspect that some sort of competition had developed among the island's youth to see who could produce the largest and ugliest specimen for the redoubtable Dr Mengden—with the arrival of each snake, and the entertainment provided by my inexpert untying and bagging of the creature, came the query, 'When is Dr Greg coming?'

Knowing next to nothing about Melanesian snakes, I was uncertain which were venomous and which not, and I was soon regretting my offer to help Greg out. One afternoon, a spectacularly enormous snake of extremely irritable temperament arrived with half the village following it. The evil-looking thing was almost three metres long, olive brown, and had a huge head and thick body. It had been tied to a stick with bushrope, and as I unleashed it from the tail up I became aware of its prodigious strength. I selected the very largest cloth bag we had, untied the last bonds holding its neck and swiftly hurled the writhing fury into the sack before tying it tightly shut. The only place I had to store it was among the rapidly multiplying canvas bags that adorned the rafters of our dwelling, but the bag was so long that as we slept the creature writhed and hissed just centimetres above our heads.

Despite such inconveniences, we soon settled into a routine, by day setting mist-nets (which are like fine fishing nets and are set on poles) in order to sample bats and cage traps for rats, then spotlighting and monitoring the nets and traps at night as we searched for other wildlife. It was an exhausting routine, but as our time on Woodlark was limited we needed to make the most of each night. On top of this workload, we needed to investigate caves in which cave-dwelling bats might live. When we'd first arrived I explained this to the head of the village council. My request for help in identifying such caves elicited an enthusiastic, if wide-eyed, response. I'd been too busy to follow up immediately, so I was mildly surprised when he arrived at our residence one morning and enquired anxiously about when we would go to the caves. I suggested that, if it suited him, we could go the following morning.

I was expecting a long walk, but to my amazement a pick-up truck—one of the few on the island—arrived at the crack of dawn.

In it sat the councillor and his wife, and soon we were rattling towards a limestone ridge a few kilometres inland. As we bumped along the dirt track, the councillor confided in a conspiratorial tone that treasure—in the form of Kula shell valuables—was to be found in the caves. One kind of Kula valuable is made of pearl shell ornamented with carved nuts and beads. These are exchanged clockwise around the eighteen islands of the Kula Ring in magnificent traditional canoes. A second kind of Kula valuable in the form of shell armbands is exchanged counter clockwise. The custodianship of the valuables brings tremendous prestige. Yet they are never held by one individual for long—it is the relationship between the traders that is valued. A Kula partnership is a bit like a marriage, creating lifelong bonds and obligations. The most esteemed Kula valuables have individual names and histories, and the most important chiefs may have hundreds of Kula partners, and ownership of the valuables, even if only short-term, is essential to status.

Still whispering, barely audible above the noise of the engine, the councillor requested that, if I happened to see any Kula valuables in my wanderings, would I mind bringing them out with me? Then he casually dropped the news that the caves had been used as a cemetery since time immemorial and were haunted. Finally he stressed that, if I found a Kula valuable, under no circumstances should I hand it to himself or his wife. Instead, I should go and sit in the back of the pick-up with the object in my hands, and when we got back to Guasopa I should place it on the ground beside the councillor's house. If it was still there the following morning, he said, then the ancestors must want him to have it and he would become a rich man. These rather mystifying instructions left me with more questions than answers and a feeling that *masolai*, as

ghosts and spirits are known in Melanesia, were believed to guard
the caves. In any case, it was dawning on me that I was being kept
in the dark about many aspects of the enterprise.

The truck stopped at the top of a low limestone slope covered
in primary rainforest. The trees were huge and liana-hung, and the
calls of birds and insects sounded everywhere from the undisturbed
bush. Clearly this was a place where the islanders didn't go. I took
a few steps away from the truck, looking back at the councillor. I
was expecting him to lead the way, but he was sitting determinedly,
hardly daring to look towards the bush much less get out of the
vehicle. Noticing my confusion, he gestured westward, telling me in
an agitated voice to go that way and to keep walking until I reached
the caves. So I set off in my new role as ghost-whisperer.

It's very easy to get lost in such country; the broken limestone
is like a labyrinth, and the bush is so dense that one can quickly
become disoriented. It took me several hours of cursing and
wandering, and avoiding half-grown-over sinkholes, to find
anything like a cave. Unfortunately the one I came across had
collapsed, but a few bones and pottery fragments among the rubble
indicated that it had indeed been used for a burial. Unfortunately,
neither bats nor Kula valuables were in evidence.

As I stumbled back to the car, flailing at insects and vines, and
covered in sweat, the eyes of the councillor and his wife widened
in alarm. Though I must have looked an uncouth sight, at first I
didn't understand the reason for their panic. Then it dawned on
me that perhaps they thought the *masolai* had, as a punishment for
my impertinent investigation, sent me stark raving mad. Feigning
the wild-eyed look and antics of a man possessed, I dashed towards
the truck, and for about two seconds I thought that the councillor
and his wife might die of fright at the sight of this devil-possessed

dim dim (white man) rapidly closing in on them. Regretting my joke, I managed to convince them that I had encountered neither Kula valuables nor ghosts, at which they seemed mighty relieved, if a trifle crestfallen.

The experience had left me pretty knocked-up and that evening, after walking the trap-lines and checking the nets, I finally crawled into my sleeping-bag around midnight. In the wee hours I was roused from a deep sleep by what sounded like a maniac loose in the house. Tins, cups and plates were being hurled around the room, people were screaming, and a huge whiplike object was swinging above my head. For a second I feared that the cave-dwelling *masolai* might indeed have followed me home, but I grabbed a torch and saw that the ruckus was caused by a metre and a half of very strong and angry snake. It had forced a hole in the corner of the canvas bag and was trying violently to free itself, a cat- or cuscus-sized bulge in its midriff the only thing preventing it from doing so.

There was no option but to lie as flat as possible and grope around the floor for an implement with which to restrain it. Grope as I might, the only thing that came to hand was a rubber thong. Thus far I'd managed to stay low enough to avoid the thrashing creature, but its enormous head was now waving about in my direction, its eyes full of indignation and anger. To this day I don't know how I did it, but somehow I managed to pin its head against the wall with the thong, then to corral the writhing mass precariously under it long enough for Tish to find another cloth bag big enough to hold it. Thrusting both snake and holed bag in, I cursed the absent Dr Mengden and his slithering objects of study, and hoped that he would not be too long delayed.

As it transpired, the great snake expert arrived the following evening. Greg was an Olympic wrestler in his youth. With a big black beard and muscular frame, he has an imposing presence, softened only by the kindest of smiling eyes. When I saw him that evening, however, those eyes were not smiling. Greg's stay among the Islands of Love had been beset by misfortune.

CHAPTER 2

Arrival of the Snake Man

I had hoped to visit Kiriwina, but the workload on Woodlark was just too great. So I asked Lester Seri to do some collecting there and in particular to keep an eye out for a large bandicoot that might still inhabit the island. I'd found a clue to its presence while rummaging in the museum collections, in the form of a stuffed skin whose tag stated that it had been collected forty years earlier on Kiriwina. Bandicoots are rabbit-sized, ground-dwelling marsupials that eat invertebrates and fruit. In the lowlands of New Guinea, a genus of bandicoots with spiny fur, *Echymipera*, tends to predominate. Bandicoots have the shortest gestation of any mammal—just eleven days in some species, and females can give birth while they are still suckling from their own mother. Such rapid reproduction is the key to their survival, for they are a major food source in many villages and without rapid replenishment hunting would lead to their extinction.

The specimen from Kiriwina differed from all others I'd seen in its large size, chocolate brown colour, and in having hefty pre-molars which seemed capable of crushing hard foods such as nuts. It was clearly a species new to science, and it had been collected by my predecessor-bar-one as curator of mammals at the Australian Museum, Ellis Le Geyt Troughton. Why, I wondered, had he not described the creature and given it a scientific name? That bandicoot would set me on a quest that would reveal as much about the history of the museum that had employed me as it would about island biodiversity.

The Australian Museum's archives revealed that Ellis Troughton had spent his entire working life at the institution. He had joined aged fifteen, in 1908, with the very first intake of museum cadets. The position of museum cadet has long since vanished, but after a few inquiries among the older staff I discovered that it was a kind of apprenticeship. The idea was a novel one for a museum, and it seems to have originated with the museum's director at the time, Robert Etheridge Jr, who served from 1895 until 1919. The museum board probably approved the idea because cadets provided cheap labour—a mere £26 per year—half an average museum salary. But there may have been another reason why they were instituted, and in particular why Troughton was chosen in the first intake.

John Calaby, one of Australia's greatest mammalogists and a keen book collector, thought he'd discovered a clue when he was trawling through a second-hand bookshop in Sydney. He came across a children's book called *The Dumpy Book of Animals*, which was inscribed, 'To Ellie, on his 9th birthday, with love from his father'. John recognised the handwriting as that of Robert Etheridge Jr, and reflecting on the striking physical similarity

between Etheridge and Troughton as revealed in old staff photos,
surmised that Troughton may have been the illegitimate son of the
director and the museum's charwoman.

Starting on the bottom rung of the museum hierarchy,
Troughton soon rose to the prestigious position of curator of
mammals, a post he held until 1957. His great work, *The Furred
Mammals of Australia*, was published in 1941 and for decades it
remained the definitive book on the subject. Indeed he was so
esteemed by his peers that he became the first life member of the
Australian Mammal Society. Today, the society's highest award—
the Troughton Medal—is named after him. Between the 1920s and
1940s Troughtie, as he was widely known, undertook a number
of heroic collecting expeditions to the Pacific Islands. In 1944 he
collected the bandicoot that intrigued me. By then he had been
seconded to the US Typhus Commission, which was set up to determine why so many soldiers were dying of the disease. Transmission
from rats and bandicoots was suspected, which explains why
Troughton, Australia's leading expert on these creatures, was
recruited to help with the inquiry.

Perhaps wartime duties, such as his responsibilities as museum
air-raid warden, prevented Troughtie from examining the bandicoot closely. Whatever the case, by the time I arrived at the museum
the remains of what was clearly a very distinctive new species had
lain in obscurity for forty years. Nothing more had been reported
of it, and one of my first jobs in my new position was to name it. I
chose *Echymipera davidi*, for my newborn son David.

Although I never met Troughtie he was a very real presence in
my life. On my first day at the Australian Museum I sat at his old
wooden desk and could make out faint doodles imprinted, perhaps,
by a young Ellie in a moment of boredom. After his lifetime

of sitting at it, the desk was probably covered in his DNA; the museum library was certainly filled with his books and its collections replete with his specimens. Fascinated by the man, I began to ask older colleagues who had worked with him what he was like. It turned out that Troughtie had never married, being described as a 'confirmed bachelor'. A colleague who had once visited him at home recalled that his personal library contained almost no works on science, but was instead crammed with books on theatre. Just occasionally his love of theatre can be glimpsed in his scientific work, such as when he gleefully describes collecting bats from a Sydney church, assisted by the rector playing a *danse macabre* on the church organ in order to keep the creatures on the wing.

Museums, like all institutions, have a rich store of anecdotes. One of his closer friends recalled that Troughtie's greatest joy was to go to the Digger's Club at Bondi to enjoy a beer and the impromptu show put on by the members, then recently back from the Pacific war. The building was a tin shed, filled with benches and tables, with a curtained stage at one end. After the men downed a few beers some Hawaiian guitar music would strike up, and the curtains would open to reveal a chorus line of out-of-form diggers hula-ing across the stage, dressed in grass skirts and sporting half-coconut shells as breasts, the denouement coming when the dancers lifted their skirts. He would also go to his local rugby league club and shout the players beers until they picked him up, tossed him in the air, and caught him again. Then Troughtie was in bliss.

Troughton never really left the museum. In a forgotten corner of that grand institution an 'old man's room' was set up, with desks for the superannuated curators, and it was to this room that Troughtie 'retired' when he left the public service in 1958. For as long as his health permitted he came to this room, as punctually as if still employed. Another of my colleagues had once gone to see him there. Troughtie was by then a frail old man, who was sitting rather awkwardly at his desk. Curious, my colleague saw that his odd posture was caused by a large cardboard box which lay between the chair he was sitting in and the desk. It was full of old shoes—perhaps every pair that the ancient curator had ever owned. After a lifetime in a museum, the impulse to collect can manifest in strange ways.

Lester did manage to collect a single specimen of *Echymipera davidi* in Kiriwina, demonstrating that it still survived on the densely populated island. The DNA sample he took from it was vital in unravelling the evolution of the group. But all had not gone well for the expeditioners on Kiriwina and, when we met on Woodlark, Lester told me the whole sorry story. Greg Mengden had spent several frustrating days on the island without seeing any interesting snakes. Then, just moments before he was due to embark on the *Sunbird* to come to Woodlark, a villager arrived carrying a small serpent. It didn't look interesting to Lester, who said it resembled a rather thick shoelace, but when Greg saw it his excitement rose to fever pitch. It was, he exclaimed, a *Toxicocalamus*, one of the most obscure of snakes, until then unknown on Kiriwina. The unexpected opportunity to photograph and study such a rare creature could be the highlight of Greg's expedition.

As the *Sunbird* was anchored in the lagoon and had to leave on the high tide, there was not a moment to lose. Greg walked to the edge of the village, placed the reptile on the sand under a coconut palm, and started snapping away. The only problem was a fly that kept landing on the snake's nose, and it was soon joined by others. Just why the winged creatures were so persistent did not become evident to Greg until he withdrew his hand from his camera, and he saw a brown smear on his fingers. A glance confirmed that the entire barrel of the lens was smeared with the same stuff, and it only took a whiff to confirm its identity.

Faeces had somehow besmirched his hands, camera and the rare snake before he had noticed, and now the flies were coming thick and fast. Wondering what diseases he had just contracted, yet unwilling to give up the splendid opportunity to photograph the rare beast, Greg washed his hands, and even took the time to give the camera and snake a quick clean. Still the flies came and, now that he'd noticed it, the odour was growing worse. Somehow, in his excitement, Greg had chosen the children's latrine as the location for his photography session. Besieged by the stink and buzzing flies he finally gave up the attempt, straightened his back, and was about to bid farewell to the assembled villagers when a young boy approached with a coconut husk in hand and said solemnly, 'Excuse me masta, there is excrement.' 'Excrement?' thought Greg, marvelling at the unexpected excellence of primary school education on the island. Then the lad bade him bend over and, before the assembled crowd—which by this stage was in gales of laughter—used the husk to spoon an enormous turd from the seat of Greg's jeans. It represented a tremendous effort by some toddler, and Greg must have sat in it when he first squatted down to shoot. His hand, which he extended to steady himself, had with unerring aim found another somewhat lesser offering, explaining the smears on the camera.

Desperation now seized our intrepid herpetologist—both to be rid of his shit-smeared clothing and to be gone from Kiriwina for good. Striding towards the beach, he waved urgently at a couple of kids sitting beside a small outrigger canoe and requested they take him to the *Sunbird*, which was lying at anchor about a hundred metres offshore in the lagoon. Greg weighed around 110 kilograms and he perched atop the tiny *lakatoi* like an elephant on a circus stool. Within moments of leaving shore the inevitable happened. Greg shifted his weight, and in response the canoe's outrigger lifted slowly from the water and described a majestic arc through the air. With the canoe overturned, Greg was deposited into the sea, desperately holding the precious snake and camera above the briny with one hand, and urgently signalling to Matt Jumelett for rescue with the other.

Matt teased Greg relentlessly. But, freshly clothed and with a cold beer in hand, even Greg himself could see the funny side of things. After all, here he was aboard a catamaran afloat in a tropical lagoon gliding over a warm sea into the setting sun. And he could always photograph his snake at Woodlark. The world seemed good again. But then the slight swell that heralded the Solomon Sea began to be felt. The *Sunbird* had slipped beyond the protection of the reef. Greg said that the beer didn't taste so good anymore, and the lingering whiff of faeces seemed to intensify. Then, Lester said, our herpetologist suddenly turned green. Not having sailed before, Greg was unaware that he rivalled the redoubtable Des Beechey in susceptibility to seasickness. He was to spend the twenty-four-hour journey in a welter of agony, either in the head (as ships' toilets are known) or lying stupified on his bunk.

Despite his sufferings on the restless sea, when he stepped ashore on Woodlark Greg immediately set to work. He'd brought his own duty-free with him—a litre of Jim Beam whiskey—and this became his consolation as he faced the hundred or so bagged snakes that awaited him. By the time he had set himself up on a deckchair in front of our accommodation, it seemed that all two thousand

Woodlark islanders were seated in front of him. The show was about to begin, and the *masta bilong snek* would not disappoint.

As the sun set, the circle of light cast by a kerosene lantern caught a mountain of a man at its centre, his face still somewhat green above his ample beard. On his left side lay a mound of snakes in bags, while on his right stood the bottle of Jim Beam and an assortment of hypodermic syringes and vials. After several mighty swigs from the bottle Greg turned his attention to the pile of snakes. The audience held its collective breath as he opened the first bag, drew out the snake and injected it with a chemical that promoted cell division in preparation for molecular sampling. For a moment it writhed wildly in Greg's hand, causing several young ladies in the front row to spring to their feet and flee. An involuntary scream then swept the audience as Greg placed the creature, unrestrained, on the ground in front of him. He needed to wait for an hour or so after injection before taking his sample. If the creature attempted to slither towards the assembled crowd, Greg would just calmly retrieve it by the tail.

Anxious to be rid of the monster that had disturbed my sleep, I presented the near-escapee to Greg, who told me that it was a brown tree snake. I had seen brown tree snakes in Australia, but there they are usually brightly banded in brown and white, and much smaller. The brown tree snakes of Woodlark, Greg explained, were unusual in reaching such a gigantic size, and in being olive-coloured. Brown tree snakes belong to the family Colubridae—venomous snakes that have their fangs located towards the back of the mouth, are bad-tempered and frequently bite. A large colubrid has a wide enough gape to get its fangs into a human hand and can deliver a potentially fatal bite. It was the only reptile that Greg returned to its bag after injecting.

The crowd watched in fascination as the Jim Beam vanished and the pile of writhing snakes grew. The mood had reached fever pitch, each near-escape drawing howls of terror and waves of laughter as the creature slithered towards one person or another, scattering all. But they rushed back when Greg grabbed each offending snake so as not to miss a second of the fun. Then came the finale, as Greg began taking the blood and venom samples. He would seek a vein, slip a hypodermic syringe into it, and draw out a small amount of bright red blood, which he then emptied into a plastic tube. To take the venom, he'd get the snake to bite the edge of a small vial. Then he'd drop both tube and vial into the liquid nitrogen cylinder. It was like a magic show, the samples vanishing with a puff of white vapour and an ominous fizzling. All too soon, the marvellous evening of wizardry and derring-do was ended. With the Jim Beam bottle almost empty and all the snakes safely bagged again for release or preservation in the morning, Greg rose from his seat and lurched off to bed.

When the *Sunbird* returned from Kiriwina Matt discovered that the outboard on its tender (an aluminium dinghy) was not running well. He decided that the problem lay with the fuel-mixture screw, and so he set about trying to adjust it. Lester, who was keen to earn some brownie points with our captain, offered to help. Matt, however, fancied himself as a top mechanic—which he may have been in his younger days—and so the offer was brushed aside. The trouble was that the fuel-mixture screw compresses a spring, and if the screw is unwound too far it can jump out. Matt sat in the tinnie as it was tossed in the waves, his glasses fogging over with perspiration as he struggled with the screw, which he might not have been able to see. But pride pushed him on, and all the while Lester hovered nearby in case Matt realised that he needed help from a younger pair of eyes.

An ever-louder stream of Dutch expletives alerted me to Matt's deteriorating mood as he struggled with the fuel-mixture screw. But it was only the curious sight of Lester's boots and clothes, delicately balanced on the tip of his bush-knife and making their way towards shore, that alerted me to the full scale of the impending explosion. I arrived on the scene just in time to see Matt undoing the screw to its full extent. It and its accompanying spring pinged into the sea. Later, Lester told me that it had already popped off twice, but, fortuitously, it had landed in the boat. Suspecting that Captain Jumelett had keel-hauled his crew for lesser crimes than watching on as the captain failed to repair an outboard, Lester had quietly stripped off, perched his boots and clothes atop his bush-knife, and swum for the safety of the beach.

In spite of such diversions we kept working. One focus of our research was to learn more about the strange cuscus discovered by Albert Meek ninety years earlier. It is unique to the Woodlark Island group, where it is known to the local people as *quadoi*. Around the size of a large cat, it is one of the most peculiar marsupials I have ever seen. *Quadoi* from the dense eastern forests are predominantly black with small white spots rather like a quoll. But those from the drier regions have patchwork coats with random white, tan and brown splotches, giving them a strong resemblance to a domestic tortoiseshell cat.

Only the spotted cuscuses of the New Guinea lowlands have a variable coat colour anything like that of the *quadoi*. But they are more regularly patterned, and the sexes are differently coloured. Like the *quadoi*, but unlike other cuscuses, spotted cuscus females are larger than males. Unravelling the family tree of the cuscus family has proved extraordinarily difficult, but today the working hypothesis is that the *quadoi* is a distant relative of the spotted cuscuses whose ancestors became isolated on Woodlark a million or more years ago. Cuscuses are great island colonisers and have spread further from New Guinea than any other marsupial, being found as far west as Sulawesi and as far east as the Solomon Islands. I can imagine the ancestral *quadoi* afloat on a raft of vegetation making the crossing to Woodlark, which then lay closer to the mainland.

The sparse early records of the *quadoi* led me to suspect that it would be rare, but it turned out to be abundant around Guasopa; we even found it in young saplings around houses and gardens. Earlier zoological visitors, it transpired, had worked on the wet and densely forested southern and western parts of the island, and in those places it is indeed uncommon. This, along with the absence of species reliant on wet habitats and the presence on Woodlark of several species of rats that prefer dry conditions, seemed to suggest that throughout much of its history the island had been a rather dry place, at least seasonally. Today it is located in a region of seasonally ample rainfall, but before drifting north it may have lain in latitudes where the dry season was long and harsh.

The Oxford team was carrying out a detailed ecological study of the *quadoi*, so we took the DNA and other samples required for our evolutionary studies and shifted focus to the other mammals. As I wondered where precisely to explore next a magnificent Kula trading canoe hove into the lagoon and was pulled up on the beach in front of the village. She had come from Alcester Island, a remote speck of land a day's sail to the southwest. The men who crewed her had a rugged, seafaring look, but they were very friendly, even a little shy. They explained that they had come to Woodlark because a young boy on Alcester had fallen out of a tree and broken his arm.

Guasopa was the nearest clinic and the only place where treatment was available. While they waited in the village for the boy to be tended to I talked to them about the animals that could be found on their island. Alcester, they told me, was home to a cuscus that sounded rather like the *quadoi*. This was exciting news, for only a single mammal had ever been recorded from Alcester—a fruit bat collected a century earlier by some now-forgotten traveller.

If the *quadoi* inhabited Alcester, it would be the second island known to support the species. From a conservation perspective this is important. If the Woodlark population of *quadoi* were ever threatened, the species would have a refuge on Alcester. And a very real threat does exist for such long-isolated island creatures—the introduction of competitors or predators from the mainland. Even today cuscuses are carried about aboard canoes—either as take-away food or for trade. If the common cuscus, which abounds on New Guinea and some islands of the Kula Ring such as Kiriwina, ever reaches Woodlark it could outcompete the *quadoi*, or spread diseases to which the *quadoi* has no immunity.

Sitting on the beach, I debated whether to cut short our stay on Woodlark and detour to Alcester. The southeast trade wind blew relentlessly, and young boys frolicked in the shallows, playing with miniature outrigger canoes. When placed in the right orientation to the wind, they shot across the surface of the sea like rockets, with groups of delighted children streaking through the water in their wake. Outrigger canoes might look simple, but they are among the most sophisticated constructions made by pre-industrial societies. As with a jumbo jet, no single person possesses all the knowledge required to manufacture one, and at times no single island could supply all of the specialised components, nor all the know-how required. Wherever they are still constructed, a traditional culture must thrive. Alcester lay on the Kula Ring, and the magnificent trading canoe that had been pulled up on the beach before me spoke of the survival of a vibrant, traditional culture there. We would cut short our stay on Woodlark, I decided, and go home via what is perhaps the most isolated and least visited island in the Solomon Sea.

CHAPTER 3

Alcester, the Lonely Isle

As we were heaving anchor on the *Sunbird* to leave Woodlark, a small outrigger canoe made its way out to us. On board was a triumphant young man holding a large goanna tied to a stick. Lester Seri—who had wanted to identify the goanna species inhabiting Woodlark Island—was delighted, and hurriedly paid the man for it. He then placed the creature, still tied to the stick, on his bunk and returned to assist Captain Jumelett with our departure. Mipi had a horror of reptiles. She'd not seen the goanna carried aboard, but when she heard that it was sailing with us she said emphatically that either the goanna went overboard, or she would. Matt finally persuaded her that it would be cruel to toss the creature into the sea so far from land, and she relented on the condition that it be killed and preserved immediately.

A downhearted Lester, who had hoped to bring the lizard alive to Port Moresby, went off to do this dismal duty, but soon emerged in a state of astonishment, saying that he could see nothing but a large goanna turd adorning his pillow. When Mipi heard that the goanna had done a runner she fled to her cabin, refusing to open the door until the monster had been searched out and ejected. Despite a thorough examination of the boat, however, the goanna remained elusive. Considering the situation, Lester stood outside Mipi's cabin and opined loudly, to nobody in particular, that it was most likely a sea-going species and had jumped overboard of its own accord and was even now swimming to land. Not entirely convinced, it was some time before Mipi emerged. But in the bustle of shipboard life the goanna was soon forgotten, and things carried on as before.

Alcester Island is high, so it's visible from a long way off. When first spotted, the green speck on the horizon seems delightful, but the closer you get the less inviting it looks. The island's basalt core is all that remains of an ancient volcano. It resembles a gigantic, angular stone, flat on top with sheer sides crusted at the base with limestone cliffs. With no fringing reef to hold them back, waves beat fiercely against the cliffs, carving them into caverns and spires. They reminded me of the redoubts of the cartoon wizards of my childhood.

Alcester's geology is revealing of its history. The island formed when a volcano rose from the sea. Presumably, it was initially cone-shaped, like Japan's Mount Fuji, but then erosion by waves steepened its sides, giving it a more block-like shape. As the magma chamber that fed the volcano cooled, it became heavy and began to sink into the sea, until it lay at sea level. The volcano's summit was then planed off by the waves, forming the plateau that exists today. Then, awesome geological forces gathered strength, thrusting the island skywards once again. This would have happened in stages. The limestone cliffs were clearly once fringing coral reefs that formed as the island paused in its ascent, but which have now been elevated high above the sea. In all probability Alcester is still on the rise.

As we sailed along the island's northern coastline we could see no sign of human habitation, but then, in an open cove towards the island's western end, we found a neat, if tiny, village. As we entered the calm waters of the cove and prepared to drop anchor, the *Sunbird* was surrounded by small outrigger canoes manned by curious women and children. The deep blue sea was so clear that we could see corals growing twenty metres or more below. But the submarine slope was only slightly less steep than the island's cliffs, making placement of the anchor—so that we were safe from drifting onto rocks or out to sea—a matter of some difficulty and precision.

The women in the canoes were dressed in traditional grass skirts, and were accompanied by naked children. This was very different from the situation on Woodlark, where everybody wore western clothes—except on every second Tuesday when the schoolchildren donned traditional garments. From what I could see of this village, which was tucked away among the coconut palms, it was entirely traditional. The evident lack of western influence made me feel a little like I'd arrived with James Cook on Tahiti. Later, the women told us that most of the men of the village had gone on a great Kula voyage, and that the few who had remained had gone to Woodlark to seek treatment for the boy with the broken arm. Alcester was thus a tropical paradise temporarily inhabited only by women and adorable, energetic children.

As idyllic as the tiny island community seemed to us, it had its problems. When two small boys climbed aboard the *Sunbird* they asked for just one thing—a glass of water. It was the height of the dry season in a very dry year. The villagers had just a few litres of water in the bottom of an old tank, which they were holding as an emergency supply. They were subsisting on the juice of coconuts and they were very thirsty. We gave them some water, but our own supplies were limited, so we could not do as much as we would have liked. The villagers were delighted, however, when we produced a huge tuna from the freezer. It was enough to feast the entire community.

I was curious to learn whether any other ships had called recently at this remote place. A woman told me that the last vessel to anchor there was a yacht hired by one of the major cigarette companies. It was, according to her, luxurious, and its crew had given away free cigarettes before screening romantic films in which the actors looked sexy, fit and powerful as they puffed on their cancer sticks. This example of modern capitalism sowing addiction and death in paradise sickened me and left me wondering just how long the traditional culture of the island might survive.

That afternoon, after setting mist-nets and rat-traps, we rested or entertained ourselves by snorkelling and diving from the *Sunbird*. I had never seen water so clear, and as I dived towards the bottom my ears crackled and popped as the pressure increased. Without a weight belt, it took most of my energy just to get down, but once there I was amazed at the abundance of tiny fish and the brilliant colours of the corals and worms. Cooled by currents sweeping up from the ocean depths, Alcester Island's corals had remained untouched by pollution or coral bleaching. Little did I realise back then that, due to climate change and coral bleaching, I'd not see its like again.

As evening drew near we went ashore to search for the island's mysterious *quadoi*. I'd been feeling slightly off colour all day, with

pains in my legs and a headache. I knew the feeling all too well.
It was the onset of malaria—a disease that had been a constant
companion ever since I'd begun work in Melanesia. By the time
we'd started the climb to the plateau I was sweating and finding
it difficult to walk. Lester and Tish suggested that I rest in a hut
in the village. Feeling frustrated and angry that my only chance to
see the island had been taken away, I tossed and turned in a bush-
materials bed in a pitch black hut, cursing my bad luck.

As I began to feel increasingly nauseous and fevered, a slight
click at the door alerted me to someone's presence. Scared and not
knowing who it might be, I turned my torch towards the door and
saw there a girl in a grass skirt. She looked to be around fourteen
years old, and to my fevered eyes appeared as if she'd just stepped
off the film set of *Mutiny on the Bounty*. Her innocent face was
framed in a halo of curly black hair and she was carrying a fan, a
bowl of precious water and a moist cloth. Without saying a word she
sat down beside me and began mopping my body. In the darkness
and silence I could hear her breathing quietly beside me, and as I
cooled the nausea retreated. How many people, I wondered, would
trust their teenage daughter to a total stranger in a darkened hut?
Yet some woman, who may not have even spoken to me, had clearly
told her daughter to sit with me. Overcome with gratefulness for
the kindness of that stranger, I eventually drifted into a deep sleep.

It was the small hours of the morning when Lester woke me
with news of his very successful night. He had obtained samples of
both *quadoi* and the island's flying fox. I was feeling much better, so
I helped Lester skin the catch with the intention of giving the meat
to the villagers, whom the drought had placed on short rations.
As I set to work, it was immediately obvious that the cuscus of
Alcester Island was indeed extremely similar to that of Woodlark.

But on Alcester, Lester reported, the creatures were exceptionally abundant. So much so, he said, that he had taken aim at a flying fox in a tree, and out of it had fallen both a bat and a *quadoi*. They must have been in line with his rifle, feeding on the same bunch of figs. Laboratory studies later revealed that Alcester's *quadoi* had most likely been introduced to the island in relatively recent times—probably in the last few thousand years—from Woodlark. Archaeological studies later suggested that such inter-island transfers had been a common practice among the people of Melanesia for thousands of years, and were presumably a deliberate strategy to supplement the meagre larder of game animals found naturally on many of the islands.

The following day Tish and I packed up the nets and traps. If the TAMS volunteers were to meet their flights in Cairns we had to depart the island by lunchtime. There was time for just one more piece of work. We had been told of a sea cave where small bats roosted, and Lester had set out by canoe to examine them. He returned just as we were preparing to up anchor, and was still somewhat shaken. The approach to the cave was perilous, and if the *lakatoi*, which was crewed by children, had capsized, the waves and jagged limestone would have turned him to mincemeat. Thankfully, the kids were expert seafarers, and he returned with news that the cave contained sheathtail bats belonging to two species. These tiny brown creatures are so named because the bony part of their tail is encased in an extensive skin membrane. They are common in Melanesia, and are often found roosting in sea caves.

When we tallied the discoveries made during our twenty-four hours on Alcester, we realised that we'd documented six mammal species. Over the previous century of mammalogical research, just

a single specimen of a flying fox had been recorded from the island.
All in all it wasn't a bad day's work.

As we left Alcester in our wake and set course for Alotau, the
capital of Milne Bay Province, my mood was tinged with sadness.
Travelling the islands of the Solomon Sea by catamaran had been
a sublime experience, but in Alotau I would have to bid adieu to
the *Sunbird* and most of my companions. We had formed genuine
bonds of friendship and had shared amazing adventures. Now they
would sail back to Cairns, while Tish, Lester and I went by air
to explore one of the most extraordinary islands in the Pacific—
Goodenough in the D'Entrecasteaux Group.

We approached Alotau and began to sort through our piles of
equipment, deciding what should follow us to Goodenough and
what should return to Australia. As Lester lifted a large box of
traps that had been stored just outside the head, a familiar figure
emerged. It was the goanna that had left its calling card on Lester's
pillow. I was amazed that a creature over a metre long could go
unnoticed for so long aboard a crowded vessel the size of the
Sunbird. It was a lesson in how easily stowaways like the Pacific rat
and the house gecko could travel unnoticed on outrigger canoes
piloted by the ancestral Polynesians, and so spread themselves to
every inhabited island in the Pacific.

We tried to keep news of the goanna's re-emergence from Mipi,
but she noticed our attempts to catch the animal and promptly
locked herself in her cabin once more. After a scramble Lester
finally grabbed the stowaway, and placed it securely in a large plastic
box. Matt then took pity on it and fed it with leftover chicken bones.
This had an astonishing effect. The hitherto wild lizard became
as tame as any lapdog, reaching up to take the bone almost from
Matt's hands. With each bone it gulped down, our captain's heart

visibly softened. Words were said about it becoming the *Sunbird*'s mascot, but much to Mipi's relief Lester informed Matt that, now that Papua New Guinea was an independent nation, it was illegal to export wildlife without a permit. Then Lester announced that he too had become fond of the reptile and could not make a specimen of it. Fed and recovered, it made an assisted leap for freedom into Alotau Harbour, its skill at swimming proof that Lester's initial identification of the beast as a mangrove monitor was correct. The species is widespread in coastal Melanesia, and it would have been quite comfortable in its new home among the Alotau mangroves.

CHAPTER 4

Goodenough Not Near Enough

So it was that Lester, Tish and I stepped off the *Sunbird* planning to survey one more island before returning home. In contrast to Woodlark, Goodenough Island is close to the mainland and accessible by a regular air service. Indeed I had seen it once previously. A few years earlier, when flying along the mountainous spine of southeast New Guinea, cloud had obscured all but the three highest peaks at the eastern end of the range. I soon identified two as Mount Suckling and Mount Dayman—the highest points on the tail of New Guinea's mountain backbone, if you imagine the island as a gigantic bird. But I struggled to identify the third peak. It was an abrupt rocky spire, lying to the north of the others. Only later did I realise that it was not part of the mainland at all, but the summit of Mount Goodenough.

Goodenough Island is the westernmost of the three islands that comprise the D'Entrecasteaux Group, and it must be one of the tallest islands for its size on the planet. All three are old fragments of continental crust that became detached from New Guinea some time between two and five million years ago, and although the strait separating the islands from the mainland is narrow, it is very deep, as are the waters separating the islands from each other. Each island in the D'Entrecasteaux Group is thus a separate experiment in evolution, and none is more intriguing than Goodenough. Zoogeographic studies show that large, high islands have a far greater chance of retaining a diverse fauna than do small, flat ones, and because Goodenough is so ancient, large and high, we felt that it might reveal species that are effectively living relics from an earlier stage of New Guinea's development.

Goodenough's mammals had been investigated on just two previous occasions. In 1896–97 the redoubtable Albert Meek had spent a few weeks there, but he'd had a terrible time of it. He desperately wanted to ascend the peak that dominates the island, but hostile locals prevented him. In that understated way that is so typical of nineteenth-century explorers, he wrote that:

> On the way up the mountain, going through the garden of a
> village, I encountered a native who threatened me with a stone
> axe and tried to turn me back. I kept going steadily forward
> though he brandished the axe in my face. He came so close
> that I feared one time that I would have to shoot him.[5]

Such hostility ensured that the peak remained *terra incognita*, and almost sixty years would pass before another mammalogist would attempt the climb. In 1953 the Fourth Archbold Expedition landed on the island and spent a month there. The Archbold Expeditions—which ran from 1933 until 1964—were tremendous affairs involving dozens of people, mostly researchers from the American Museum of Natural History, who could be away from home for up to eighteen months at a stretch. Financed by the millionaire philanthropist Richard Archbold, who participated in the heroic expeditions in the 1930s, they represented the most concerted effort ever to document the fauna and flora of Melanesia. Often using seaplanes, the early expeditions specialised in penetrating unknown territory, making first contact with tribal people, and discovering dozens of new species of birds, mammals, reptiles and plants.

The 1953 expedition was a sedate affair compared with its predecessors. It visited regions made newly accessible by the frenzy of airstrip construction that accompanied World War II. After collecting in the lowlands of Goodenough, the mammalogist on the expedition, Hobart Van Deusen, ascended the mountain and camped for around two weeks on a forest ridge-top at 1600 metres elevation. In the official expedition report he noted that:

> Results in the mammal department would have been very poor, but for the success of a Garuwata man named Vilaubala who, with a small boy as helper, spent 11 days with us and hunted in the forest with dogs. Secured only this way were a small black wallaby...a black-headed bandicoot and *Dobsonia* [a fruit bat].[6]

The wallaby caught by Vilaubala was like nothing known from anywhere else, and Van Deusen named it *Dorcopsis atrata*, the black gazelle-headed wallaby. Five-million-year-old fossils of similar creatures have been found near Waikerie in South Australia, but today gazelle-headed wallabies survive elsewhere only in the lowlands of New Guinea. So why was this species found only in the mountain forests of Goodenough? It seems likely that vast geological movements stranded its ancestors there when the island separated from the mainland millions of years ago. Perhaps, we speculated, Goodenough's mountain forests were a relic of the original habitat of the genus. This was enough incentive for a visit. But we also had secret hopes. If such a large and mysterious creature could remain hidden on Goodenough Island until 1953, what else might lurk there?

A year or two previously I had travelled to New York to examine Van Deusen's specimens, which are all housed in the American Museum of Natural History. It's an extraordinary institution, holding what is arguably the world's largest collection of biological specimens. At the time of my visit, travelling scholars were given extraordinary freedom. I was even given my own key to the collections so that I could come and go as I pleased—a welcome change from the strict regimens of European museums. Moreover, the mammal collection was splendidly curated and laid out, making it easy to find and examine the specimens. I soon realised, however, that no tissues suitable for DNA analysis or skeletons of the wallaby had been collected by Van Deusen. And, extraordinarily, the collection did not include a single female. This meant that nothing at all was known of the species' reproduction. More collecting was essential if fundamental questions were to be answered.

The opportunity to visit New York had come about in an unusual way. A man who worked in a jewellery store in Manhattan had read of my work in a newspaper and posted me a cheque for $1500 to help. I decided to use some of this generous donation to go to New York, and wrote saying that I'd like to thank my benefactor in person. It turned out that his name was Eric Fruhstorfer, and he worked at Van Cleef & Arpels. He mentioned that the store was holding a party during my visit, and that I was welcome to come along. An extremely tight budget saw me staying at the YMCA, in a tiny, stiflingly hot room that stank of stale urine. Knowing that appearances counted, I carefully conserved my sole respectable shirt for the night of the party, but alas discovered that I'd forgotten to pack any trousers other than blue jeans.

Eric had given me the address on 5th Avenue by phone, but I walked past the entrance several times before realising where I was. I'd been looking for a glass-fronted jewellery store with chains and rings in the window, not an elegant granite facade and a doorman. Eric greeted me, explaining that the party was to honour everyone who'd spent more than a million dollars in the store that year. This alarmed me, and it was only Eric's extraordinarily warm welcome that prevented me backing out the door. He ushered me in as if I was the most important person there. Stunned at the unexpected turn of events I sipped my French champagne, munched on the caviar and took in the scene.

The gathering was elegance personified; the room was filled with the most beautiful women and suave-looking men, who, I was hugely relieved to discover, didn't notice me at all. Then I saw that the wait staff were converging on an elderly gent who had on his arm the most elegant blonde woman I'd ever seen. His trousers had somehow come loose and were around his ankles— a fact he seemed oblivious to. I looked on in astonishment as the waiters formed a human shield while a waitress gingerly raised the trousers and secured them. I had, I concluded, stumbled into a world more eccentric and singular than anything that exists in Melanesia. Indeed it was beginning to make stone-age cannibals look rather ordinary.

Then Eric introduced me to a tiny black man who was surrounded by exquisitely beautiful girls, most of whom looked

like they should still be in school. He was, Eric said, Baby Doc's
Minister of Art. When Eric had visited him in Haiti earlier in the
year, he had carried $9 million worth of jewels in his briefcase.
When he arrived at the mansion he found the minister in a dress-
ing gown on his bed, surrounded by eleven girls in various states
of undress. 'Come and meet my god-daughters,' he'd said as he
invited Eric in. I began to feel that I'd travelled quite a distance
since leaving the YMCA.

I recalled this strange encounter as we flew into Goodenough. The
airstrip is located on a parched, kunai-covered plain that lies in a
rain-shadow on the northern side of the island. It was constructed
during World War II and had served as a major base for allied
bombers that were harassing the Japanese. Many damaged aircraft
were abandoned there, and at the time I visited their remains were
so abundant that several village houses had been built entirely of
aircraft aluminium, some of it still bearing American insignias and
other wartime markings. Another curious legacy of the war was
that there were no traditional wooden spears to be seen. Instead
metal spears, fashioned from war *remanié* and used for pig and
wallaby hunting, were ubiquitous.

   The drought that was ravaging Woodlark and Alcester had
not bypassed Goodenough. Indeed, it had been exceptionally dry
there for almost a year. To make matters worse there had been a
mass die-off of fish in the ocean around the island, perhaps due to
volcanic activity or a red tide (a toxic algal bloom), leading to a criti-
cal shortage of food. By the time we arrived the dry season was in
full swing and the mango trees were bearing immature, sour green
fruit about the size of apricots. As far as we could see that was
the only food that remained at all plentiful and the famished

villagers had turned to eating them wholesale, so decimating the mango crop that they would otherwise have enjoyed around Christmas time.

The dried kunai grass was ablaze on the slopes all around the landing strip as our aircraft touched down. The pall of acrid smoke gave a rather funereal air to the place. Nobody came to meet us, so we walked up the hill to the nearest village to seek out the head of the village council. We found him resting, listless, on the porch of his house. When we explained that we wished to climb the mountain in the centre of the island to search for animals, he assigned three young boys to guide us. The forest edge, they said, was no more than an hour and a half's walk away. We began to prepare for an immediate departure, hoping to camp at the forest edge by lunchtime and to press on towards the peak the following day.

There is something utterly oppressive in climbing a steep slope covered in recently burned tropical grass. The heat is ferocious, and ash rises in miniature willy-willies to choke the passerby. After only a short distance we were flushed and perspiring, our throats on fire. It didn't help that our last night in Alotau hadn't been entirely abstemious. Lester had run across some buddies who had insisted on a session of beer and billiards. Tish and I joined them briefly but, aware of our dawn appointment at the airport, we'd departed around midnight. On seeing Lester the next morning I doubted whether he'd been to bed at all. Now, with the kunai-covered ridges towering over us the SP Brownies (Papua New Guinea's favourite beer) were beginning to tell, and we all cursed the merciless sun beating down upon our sorry heads.

From our vantage point it looked as if a climb of around six hundred metres would bring us to the summit of the ridge and

so to the edge of the forest. It wasn't a great distance, but the slope
was steep and we struggled under heavy loads of gear. The sun rose
ever higher in the sky, and by the time we neared the crest we were
suffering terribly. We held on by telling each other how delightful it
would be to complete our walk through a shaded forest glade and
to sip from the clear stream that no doubt flowed beneath the trees.

When we finally crested the rise, in near exhaustion, we
were appalled to see that ahead lay not a cool forest, but another
burned grassy slope which, if anything, was even steeper and
higher than the one we'd just ascended! Lester, whose hangover
must have been a sore trial, beat the ground in frustration and
cursed Goodenough's inhabitants, saying that the Department of
Environment and Conservation should take the lot of them to court
for the senseless burning and destruction of their forest.

By now we'd been walking for several hours, and ahead lay
a climb of at least a thousand metres. I turned to our guides and
asked them, *really*, how far was it to the forest edge? To my astonishment they answered that they didn't know. None of them had
been up the mountain before. Their uncle, in the hope that we'd
feed them, had ordered the boys to accompany us. In situations
like this, there's just one thing to do—boil the billy. As we took a
break and sipped our tea, we eyed off that blackened kunai slope.
The heat haze rising from it looked like it was coming off an oven.

It was almost dark when we made it to the top of that second
ridge, only to discover that beyond it lay yet another kunai-covered
slope, which had also been recently burned! This one, however,
was shorter, and we could see the edge of the forest just beyond it.
There was nothing for it but to camp where we were. Thankfully,
nearby we found a very narrow shelter at the base of a cliff—more
of a nook than anything else—just wide enough to accommodate

a row of sleeping people. Before collapsing in exhaustion, I tried to work out where we might be on our map. I put our elevation at around 1500 metres. With our limited water supply, it had been a hell of a walk.

To make matters worse, that night a drizzling rain set in—not enough to supply a drink, but sufficient to penetrate the shelter and thoroughly soak us. We were all up before dawn, wet, sore and miserable, and ready to tackle the last of the kunai. Though initially steep, the walk was bearable in the coolness of the dawn. Then, as we crested that final ridge, the trek turned into a pleasant downhill stroll, and within two hours we'd entered the forest where a small creek provided a much-needed drink. Soon after, we arrived at a campsite that was clearly used from time to time by the local people. Situated in beautiful, primary forest at around 1300 metres elevation, it was a perfect base for our investigations.

The campsite was the most extraordinary I've ever used. House-sized granite boulders lay strewn about like pebbles, making the place feel like a land of the giants. The shelter we occupied lay underneath a particularly gigantic one, which seemed to be the size of a church, propped up by three smaller rocks at the rear and sides. Underneath it was a flat, comfortable surface the size of a small house. Most of it was too low to stand up in, and the best area for sleeping had a particularly low roof—the rock hung just a few centimetres from our faces. Despite the comfort provided by this extensive, flat and dry living area, there was something disturbing about it. Whenever I lay on my sleeping mat I had an ominous awareness of the enormous weight poised above me. And the rocks holding it looked feeble. I worried that the smallest earth tremor might dislodge it, crushing us to atoms.

Just outside our boulder camp flowed a frigid, crystal-clear stream. Both above and below the camp it cascaded over yet more boulders to form a series of waterfalls and deep pools. I later learned that the local people called the camp and the river *Boitutudiadobodobona*. The Goodenough language, which seemed to be full of such lengthy and convoluted words, was one that I never even attempted to learn.

The forest around the camp was very distinctive. The trees were gnarled, relatively short and covered with a long, wispy, pale-green moss that I'd not seen elsewhere. It resembled the Spanish moss of the American south, and it moved in the slightest breeze, giving the place a fairytale feeling. One morning, as I sat beside the brook brushing my teeth, an extraordinary creature appeared from behind a curtain of moss. It was a magnificent brown-and-white brahminy kite, a bird resembling a small eagle. It swept silently downstream, following the torrent as it hunted for lizards and frogs. So narrow and hemmed in was its flyway that the raptor was forced to pass within a metre or two of me, and I got to look it straight in the eye as it glided past. The kite must have seen me, yet it displayed no fear—perhaps because it realised that there was nothing it could do except pass as unobtrusively as circumstances permitted.

We soon discovered that evenings on Mount Goodenough were almost invariably marked by drizzling rain, while the mornings were often bright and beautiful. It soon became my favourite ritual to rise with the first of the light and enjoy a mug of coffee while I watched wisps of smoke from our campfire curl through the crisp air and disappear into the mossy crowns of the trees. On clear mornings you could see the sun striking the bare rock and grassland of the island's remarkable summit more than 1000 metres above. Regrettably, a lack of time and food meant that we were unable to visit that intriguing habitat, so we decided to focus our efforts on conducting a thorough survey of the forested zone. Perhaps some future mammalogist will discover what lives in the frigid grasslands on that isolated peak, for as yet no expedition has ventured so far.

Blue-breasted pittas abounded in the forest, their bright red-and-blue plumage surreal against the jungle green. And small snakes—which we eventually identified as a species of *Aspidomorphus*, a 'venomous but not dangerous species' according to the reptile guide—were daily visitors at the camp. One of the first sounds we heard upon arriving was a powerful, rolling, almost moan-like call that seemed to go on forever. Although not unpleasant, along with the wispy moss and great boulders the low, haunting sound lent a sombre air to the place.

It was some days before I discovered where the sound came from. Looking into the gnarled branches I saw a sharp, blood-red eye peering down at me. It belonged to a curl-crested manucode, a relative of the birds of paradise, which is found only in the D'Entrecasteauxs and nearby islands. Predominantly blue-black, it looks somewhat like a large, iridescent crow, but its red eye and head gear of crisply curled feathers immediately set it apart. I observed it for some time as it sat on its perch. It would often cock its head at me quizzically, with more than a passing resemblance to Groucho Marx.

The curl-crested manucode's call is produced by an extraordinary trachea, or wind-pipe. During our stay, a boy shot one with his bow and arrow. Before assigning it to the cooking pot, he allowed me to take the skin for the museum's bird collection, and as I worked I was amazed to find that its trachea was longer than the bird itself. It lay neatly coiled over its breast muscle, looking for all the world like some terrible parasitic worm.

We had laid lines of box traps and strung up our mist-nets. These nets, which are around ten metres long, are made of fine nylon. Hoisted between two poles, they have five horizontal strings running their length, so that the nylon forms loose pockets below them. If a bat strikes the net it tumbles into one of the pockets, from which it has difficulty escaping. It is unharmed by the experience, and can be released or sampled as required.

Wallabies cannot be sampled by any of the techniques we had, so although we saw abundant signs of the black gazelle-faced wallaby, including tracks and droppings, in the immediate vicinity of the camp, I couldn't imagine how we'd ever get a sample. Even seeing one seemed unlikely, for scrambling over the boulder-strewn terrain was tedious, slow work, and wallabies are alert, nervous creatures.

On our first morning, however, a surprise discovery provided us with a sample of sorts. Tish had gone to wash at the stream that ran by the campsite and had noticed some bones lying in a deep pool. They were clearly wallaby bones, left most likely by a hunter who had used the camp only a few days earlier. As we looked at them lying three metres down in the pool of frigid water, the question arose as to who would dive in and retrieve them. If there's one thing I hate it's cold water, so I was making only very reluctant moves in that direction. Tish, however, had grown up in Scotland and had no such qualms. She quickly discarded her outer clothing and slipped into the freezing pool. Eventually she surfaced, goose-pimpled and shivering, but triumphantly holding the larger bones—including a skull. As she took a place by the camp fire, the young Goodenough Islanders who accompanied us stared at her in disbelief, as if they regarded immersion in the frigid stream as tantamount to suicide.

Such a sample, while good to have, is of limited value for the evolutionary studies we hoped to conduct. After several days at the boulder camp I had almost given up hope that we might do better when a dog walked casually into our shelter. Within half an hour it was followed by another, then a slight, grey-haired man with one arm strode silently up to our fire. His name, he said, was Agevagu, and he had come to help us catch wallabies. The boys at the camp were excited, telling us that Agevagu had the power to 'call up' wallabies. In times of food shortage, like the present, his magical powers over the creatures were much appreciated.

Despite lacking an arm Agevagu clearly had something on his side when it came to wallaby-catching, and over the next few days he and his dogs brought the creatures in thick and fast. Doubtless the weather helped, for during the drier part of the year the walla-bies congregate near the streams. Whatever his secret was, Agevagu was keeping it to himself, as the young people at the camp had caught nothing. Indeed, the boys told us that young people hardly ever came to the mountain anymore. It was too much hard work.

The black gazelle-faced wallaby is a graceful creature with large, expressive eyes, short ears and a rather long, elegant snout. Its fur is shiny black and rather coarse on the neck, but as soft as silk elsewhere. When you ruffle it, a pure white underfur is revealed, to striking effect. Strangely, on some individuals one or both front paws are pure white. We speculated that white paws might be useful for signalling in the dense, dark forest, or perhaps as identifying marks of individual wallabies. If so, it's likely that the wallabies have a well-developed social structure, for only animals with well-developed social structures need to recognise each other individually at a distance.

One of the wallaby's oddest features was the claws of the hindfeet, which were worn down to blunt stumps—something I'd not seen before. This might indicate a rock-dwelling habit, though the rock wallabies of Australia (which specialise in living among cliffs and boulders) have very different kinds of claws: they're so small and the footpad so elongated that they're protected from wear behind gripping toepads.

The black gazelle-faced wallaby's tail is also curious. As with the other gazelle-faced wallabies, the tail is held in an curve so that only the tip, which bears a hairless, cornified nubbin, touches the ground. I'm uncertain why this is so, but it has been suggested that it brings less of the sensitive tail into potential contact with leeches, which abound in New Guinea's forests.

One evening Agevagu let it slip that thirty-four years earlier he had helped Hobart Van Deusen of the Archbold Expedition to obtain his specimens—the first seen outside Goodenough Island. He must have been the boy Van Deusen had seen with the older hunter. Perhaps the wallaby-calling magic and knowledge of the environment and the wallaby's habits had been passed down through the generations. It is highly likely that Agevagu's dogs were descended from those that caught Van Deusen's wallabies, for good hunting dogs are essential to a successful hunt and their bloodlines are carefully preserved.

Agevagu had me thinking about what makes an effective hunter. Dogs are important, as is an intimate knowledge of the environment and the habits of the prey. But what of the supposedly mystical power used to call up the creatures? It is activated by performing a ritual, and while I had not seen Agevgu's hunting ritual I had seen similar rituals performed elsewhere in Melanesia. They're complex, often involving special smoke and foods, and they serve to bond dog and hunter, as well as to communicate to the dogs that a hunt is imminent. It may not be mystical powers at work, but it's easy to see how such rituals can increase the chance of a successful hunt.

Today, the total habitat of the black wallaby is no more than one hundred square kilometres—all at elevations between 1000 and 1800 metres on the upper slopes of Goodenough Island. Thus, it has one of the smallest distributions of any kangaroo species, making it potentially vulnerable to extinction. Furthermore, females have only one young at a time. Only one of the five we examined was carrying a pouch young, which suggests a relatively low rate of reproduction. Partly as a result of these discoveries, the International Union for the Conservation of Nature has classified the black gazelle-faced wallaby as critically endangered—just one step away from extinction. Even though it remains common in a tiny area, fire is nibbling away at its forest habitat and climate change may also be a threat. But one big change is working in the wallaby's favour: young people rarely climb the mountain these days, and this is easing hunting pressure on it. I don't know how I feel about this. So much culture and knowledge is tied up with traditional hunting that its loss is surely to be regretted.

Agevagu had brought his wife and some other women along with him, providing some welcome female company for Tish, as well as a little local culinary expertise. They cooked the wallaby meat, wrapped in leaves and spiced with mountain herbs and fiddle-heads of ferns, in traditional stone ovens. At first I was uncomfortable eating the meat of such a rare creature. But our provisions were precariously low, and it was so delicious that I quickly set all qualms aside. Although we could have used its meat, we preserved one wallaby whole in formaldehyde. It is the only specimen of its kind in the world and, nearly a quarter of a century after we collected it, it yielded a remarkable secret. In

2010 researchers published a report on their examination of some of its stomach contents. Viewed under the microscope, the stomach contents were seen to contain the spores of a diversity of fungi, including truffles. Evidently the black gazelle-faced wallaby digs up and consumes these fungi, so spreading their spores throughout the forest. Truffles and other fungi are important parts of the forest ecosystem, without which much biodiversity cannot survive. So the wallabies are vital to the forest's health.

While camped at *Boitutudiadobodobona* we encountered other strange creatures. One, a bandicoot of which we obtained just a single specimen, eludes identification to this day. Because Kiriwina was joined to Goodenough 20,000 years ago when sea levels were lower, I had hoped to find *Echymipera davidi* there. It, however, appeared to be absent, being replaced on the mountain by a bandicoot species that differed in size, colour and dentition. I had not found the time to describe and name this species before moving on from the Australian Museum. Perhaps I'm destined to play Troughton to some as yet unborn mammalogist, who will be inspired by the creature I collected to visit Goodenough Island or search the world's museums for more samples so that it can be properly described.

More common in the mountain forests was a reddish-grey tree-dwelling rodent known as Forbes's tree-mouse. They are fluffy-furred creatures that spend their days sleeping in densely packed family groups in tree-hollows high in the canopy, and sally forth at night to eat fruit, buds and leaves. The entrances to their nests are difficult to detect, and it was only with the assistance of the sharp-eyed local boys that I saw them. Sometimes the hollows are made in the vertical trunk of a vigorous young forest tree. The rodents somehow gnaw their way into the sheer trunk, leaving only a tiny entrance. Inside, the hollow widens quickly into a large cavity which is packed with leaves, moss and other bedding and which is soaking wet—something the tree-mice are clearly inured to.

These rodents have an interesting family structure. We invariably found the nest-hollow to be occupied by a pair, which was sometimes accompanied by what looked to be two generations of young, the older of which was adult-sized. The tree-mice never attempted to bite, and it may be that these gentlest of rodents form lasting pair bonds and strong family ties, allowing offspring to remain with their parents into adulthood.

As the days went on the chronic lack of food at the boulder camp was forcing us to become inventive. One evening a youth proudly carried a hornbill into camp. Normally hunted for their spectacular beaks, which are used as body ornaments, the hornbill is a tough old fowl at the best of times. Tish, however, offered to try to turn it into hornbill-a-leekie soup. Perhaps it was testimony to our starving state, but Lester and I still remember Tish's creation as the best soup we had ever dented a tooth on!

Towards the end of our stay, when our rations had completely run out, Lester and I were astonished to see Tish produce a tin from the depths of her pack. It was haggis, and Tish said that it was an early St Andrew's Day present. Despite having reservations about the Scots national dish, we scoffed it down. But finally, sheer hunger forced a retreat from boulder camp. We were all very sorry to go, for we had made the place comfortable and we would have liked more time to survey the fauna of Goodenough's high forest.

When we arrived back in the lowlands we found a tense and unhappy situation. Hunger and the seemingly endless drought were fraying everyone's nerves, but now a grave social upset had added a volatile element to the mix, the evidence of which assaulted our noses the moment we entered the village. The night before our departure for the mountain a man had died. Now, over a week later, a platform had been erected on the edge of the village, atop which lay the bloated corpse. Placing the deceased on such a burial platform is part of the traditional funerary practices in the area, but the body is supposed to be buried before it becomes offensive. In this case the deceased was a senior and respected man, but nevertheless he remained unburied due to a deep disgrace in his family. The councillor explained that the deceased's eldest son had failed to turn up for the funeral, and without him the burial could not take place.

The son had been sent to Port Moresby some months earlier, entrusted with the community's savings to buy a new fishing boat. The villagers had waited and waited, but neither man, money nor boat had ever shown up. The stress and disgrace of being related to such a notorious thief had driven one of his brothers insane, and his daughter had become very ill—perhaps, it was said, due to *sanguma* (witchcraft) wrought by a vengeful neighbour. Then, to top off the family's misfortune, the father had died.

We boarded the aircraft that was to take us back to Alotau with the smell of decay heavy in the air. The corpse was being kept above ground in an attempt to force the prodigal son to return to the village. In this test of wills being played out around the corpse I wondered who would buckle first. Just as we were leaving, there had been a move by part of the community to bury the corpse. It had been frustrated, however, and the man who waved us off predicted that fighting would erupt as a consequence. 'Like sitting on a powder-keg' was how I described, in my diary, the tension I felt while awaiting the flight.

Arriving back at the Australian Museum we had enough

materials to write descriptions of the mammal faunas of the islands we had visited. This would provide useful information for the conservation of the species that occurred there. But our ambitions were greater than that. If we were to understand the patterns of zoogeography that prevailed in the region, we would need to visit many, many more islands. Planning for those expeditions occupied much of my time in the following year.

CHAPTER 5

Island of Lepers

The *Sunbird* had proved an ideal platform for island research, and the following year she set sail again from Cairns, this time as part of a far more ambitious research program. She would cruise as far north as the west coast of New Britain, dropping scientists on Normanby Island in the D'Entrecasteaux Group, and Sideia Island near Samarai, on the way. That way we could cover four islands in a single field season. Regrettably, I had other fieldwork commitments, and so command of the complicated expedition was left to Tish Ennis, who was now a highly competent field biologist and expedition organiser. She set out with Lester Seri, who had flown from Port Moresby to Alotau to join the researchers coming from Sydney.

The idea was to drop an experienced researcher, or a small team, on each island as the *Sunbird* travelled north, and to pick them up on the return journey. George Hangay, the Australian Museum's taxidermist and an experienced field collector, was to disembark on Sideia Island, which lies just east of Samarai. We could find no records at all of any mammal collected on Sideia, yet in terms of zoogeography it was a very significant place. At just eighty-nine square kilometres it lies immediately east of the New Guinea mainland, which it was almost certainly part of when sea levels were lower 20,000 years ago. Back then it must have supported all the mammal species found in the lowlands of New Guinea. The question I wanted to answer was how well various species had fared. Had any, for example, become extinct as a result of their isolation on such a small landmass? Such studies are central to the science of zoogeography. They are particularly useful for designing national parks and reserves, which in effect become islands in a sea of land modified by humans. If the parks or islands are too small, such studies show that it's unlikely that certain species can survive. But how big does a patch of New Guinean rainforest have to be to preserve its fauna? Sideia might just help us answer that question.

As many of the specimens in the mammal section require the attention of a taxidermist, I worked closely with George Hangay. I'd developed a great respect for his abilities, as well as a fondness for the man himself. George told me that he'd been stuffing things ever since a passion for the art came to him at age twelve. A Hungarian by birth, he'd travelled widely as a biological collector, having made trips to the New Guinea highlands, Borneo and other places. With a strong, muscular physique, a dense black beard, hooked nose and dark, deep-set eyes that glowed like coals, George cut an impressive figure. I recall one evening dropping by and peering through an antique window while he was at work. There he stood in his dimly-lit workspace—which was cluttered with bones, models of organs and fabricated dinosaur heads—wearing an apron and perched over a steaming vat containing the skeleton of some large creature. His assistant was peering respectfully into the mixture. For a moment they looked for all the world like a pair of warlocks intent on concocting some fearsome brew.

Life at the museum was not always smooth sailing for George. He took on a number of after-hours jobs that some people in the museum frowned on, despite the fact that they were done at no cost to the institution. He was once commissioned to make a full body cast of a woman. She was in her thirties, and wanted her shape immortalised while at its peak. But it was the little jobs that proved most troublesome. One day an old lady arrived at the museum door in tears, asking to see the taxidermist. Her pet budgie Cyril, who was her sole companion, had died, and she carried the corpse in her handbag. 'Could the museum have it stuffed?' she asked. George was notoriously soft-hearted, and he agreed to do the job at a minimal cost. It was the beginning of a long weekend, and in his rush to get home George left Cyril on the taxidermy bench, rather than in the freezer. When he returned to work on the next Tuesday, the corpse was completely rotten.

In the thick of a busy week George pushed this seemingly small misfortune to the back of his mind. But then the old lady returned. Playing for time, George explained that the job of stuffing such a tiny creature was turning out to be rather complex and difficult. Not impressed, and perhaps sensing his evasion, the woman threatened to speak to the museum director. George was horrified. What if the director thought he was using museum materials for his own gain? That afternoon he scoured pet shop after pet shop for a budgie that looked like Cyril. But Cyril had a rather unique pattern of feathers, and George could find none resembling him.

When the phone next rang and George heard it was the old

lady waiting at the front desk for him, he burst into a sweat. Walking slowly to meet her, he had no idea what to say. He had almost decided to tell the truth and face the consequences, when a flash of inspiration came to him. With a downcast look he said that as he was skinning the poor demised creature, he had noticed certain signs of ill health. Concerned that whatever had killed the bird might be infectious, he had called the quarantine service, and to his horror the officer had confirmed that Cyril had died of *the plague*! Despite George's protests, the quarantine officer had taken little corpse away to be incinerated at a special facility. Feeling rather guilty later, he made a small wooden coffin and placed some ashes from his barbecue in it. These he gave to the old lady.

Next to taxidermy and beetles (a group he is a world authority in), George's greatest passion was wrestling. Not the Olympic type enjoyed by Greg Mengden, but the glitzy, televised, world-championship style. Indeed, George was so enthusiastic about it that he had purchased his own wrestling ring, which he would set up in outer-suburban shopping malls. Then, he would emerge as 'gory George', a baddie wrestler, and would tangle with his friend and fellow wrestler, Attila, a bodybuilder. As George explained it, the only problem was that Attila was the timidest person you could find. His nerves were so bad that he occasionally threw up in the ring, which was particularly unfortunate because he habitually swallowed twenty to twenty-five raw eggs, sometimes more than once a day, in order to maintain muscle mass.

Sideia Island appeared to be the perfect place for George—a tropical paradise in which both beetles and mammals abounded. We all expected him to return relaxed and happy, with notebooks full of biological insights. But when, after several weeks, the *Sunbird* called in on its return journey to pick him up, George was nowhere to be found. I was by this time back at the museum, where I received a phone call from a deeply worried Tish, reporting that George had vanished and that nobody on Sideia seemed to know what had become of him. For several days I feared the worst, and was plagued by dreams that George had been devoured by crocodiles or abducted by raskols. To our great relief, however, he turned up at the Australian Museum a day or two later, well in advance of the *Sunbird*. Over a cup of tea he handed me his notebooks, which were meticulously kept and full of important information, and told me what had happened.

Initially he was delighted by Sideia, and had set up camp a little distance from the main settlement. Over a week or so he collected evidence that eleven species of mammals were native to the island, of which six were bats. He was also able to determine through interviews with islanders that several mainland species were unknown there. I was delighted that he had completed a reasonably thorough survey of the island, as we could now estimate the extinction rate of mammals since Sideia's isolation. Even at first glance, I could tell from George's data that the rate must have been high, for an equivalent survey on the adjacent New Guinea mainland would have yielded twice or three times the number of species. George too, was happy, as the limited mammal fauna gave him time to pursue his own interests, such as collecting beetles and, I was surprised to learn, preserving hundreds of the enormous cane toads that roamed the island.

While on Sideia, George told me that he had dined like a king. Each evening a woman from the village would bring him delicious cooked mud crabs that she would break up with her own hands to save him the trouble. Despite the idyllic setting and delicious food,

George's field notes reveal that even at this stage he was not enjoying complete tranquility of mind. 'This is not a healthy place,' he had written, perhaps in response to the large number of islanders who seemed to be suffering afflictions of the skin and limbs. Even his cook was missing a few fingers, but then that's a common sight in New Guinea, where women often cut off digits as part of the mourning ritual for close relatives. But she was, he had assured himself, exemplary in her cleanliness.

Perhaps it was the crabs, but something was having an ill-effect on George's constitution. The villagers told him that a clinic was held each Sunday morning at a mission nearby, so he went there in search of a treatment. To his astonishment he discovered half the island's population lined up in front of a table set up outside the clinic doors. On the table sat a large glass jar, full of pills labelled 'lerpesy'. 'What is this lerpesy?' George wondered aloud as he sought something to settle his tummy. Then he remembered his cook, with her friendly smile and stumps of fingers, cracking open the mud crab claws by the light of his camp fire. With dawning horror George realised Sideia was a leper colony. He had been living for weeks, unaware, among lepers!

A small insight into the discomfort the discovery caused can be found in George's notebook: 'The leper colony has been closed down and the lepers disbanded everywhere. One should be careful staying in the villages and accepting local food.' He feared, he confessed to me, that he might already have contracted the dreaded illness. Whatever the case, he felt that there was not a moment to lose. So he packed up his materials and collections there and then, and hired a canoe to carry him to the nearest island—anywhere but Sideia.

Then, in the midst of all this activity, George found that he urgently needed to 'spend a penny'. The only toilet in the vicinity was a tiny wooden cubicle perched precariously atop sticks in a mangrove swamp at the end of a long, rickety and perilous walkway. Perhaps George was thinking again of the mud crabs as he strode along the walkway towards the box. Whatever the case, he wasn't paying attention to the rotten boards, for when he had almost reached his easement he heard an awful crack. The frail wooden structure, which was never built to support the weight of a wrestler, was giving way, and George found himself plunging from a great height into a stinking mass of mangrove mud and human faeces. As he put it to me, 'I vas plunged into de lepers' toilet up to de neck, and had to be pulled vid a rope through de stinking mess in order to reach solid ground!'

This final exposure to the awful leprosy bacterium was just too much for our intrepid collector. He set out, still somewhat bespattered and in an excited state of mind, and almost immediately found himself in further trouble. To make matters worse, he had encountered some nasty ants earlier in the day, some of which had entered his ear. He sprayed Mortein into the orifice, but obviously not quite enough because, he told me, 'One of the bloody things came alive in the middle of the passage between the two islands.'

The currents in China Strait are strong and George's canoe was overloaded. As hard as he rowed with his one paddle, he failed to make headway. In fact he found himself being carried by currents away from Samarai, and with darkness rapidly falling he was beginning to despair. For hours he was at the mercy of the tides, but towards midnight they slackened, and he saw a tiny light in the distance. He decided to paddle towards it, and soon realised that it was the glow of a cigarette. Someone was enjoying a solitary fag on a beach on an island adjacent to Sideia, and it was this that saved George from his nightmare. But it was not quite over. He stepped out of the canoe right onto a huge saltwater crocodile, and as he leapt back into the canoe he nearly capsized it, much to the entertainment of the local on shore. After resting, he found his way to Port Moresby and thence by air to Sydney.

The barrels of pickled cane toads that George had collected only made sense to me several months later when I saw the final result of his efforts. The toads—stuffed and dressed in eighteenth-century costumes stitched by his wife—had been assembled into miniature string quartets and baroque orchestras. They stood on tiptoes or sat in chairs, grasping their tiny violins and trumpets with great delicacy. George had even managed to imbue their limbs with a sense of movement and to express on their faces the concentration and passion seen in the best human musicians. Altogether the assemblages were a triumph of taxidermy—though admittedly one I hardly expected to see when I had waved our expeditioners off.

Compared with George, the other expeditioners on the 1988 voyage of the *Sunbird* had dull times. Pavel German and Lester Seri had been dropped at Normanby Island which, with Goodenough and Fergusson, is part of the D'Entrecasteaux Group. It is remarkable as the only island in all of Melanesia that has its own species of carnivorous marsupial. The Normanby dasyure is a relative of Australia's antechinus, and is known from a single individual collected by the Archbold Expedition forty years earlier. I was hoping to learn more about it, but unfortunately it had eluded our expeditioners.

Pavel and Lester did, however, make a discovery that was almost as exciting. The D'Entrecasteaux islands are home to a beautiful tree-mouse with a rich reddish coat and a long, prehensile tail. Almost as rare as the dasyure, it was known from just two individuals—one each from Goodenough and Fergusson islands. Pavel managed to spotlight and capture a specimen on Normanby, thereby demonstrating the presence of this beautiful creature on all three islands of the D'Entrecasteaux Group. Apart from a few brief notes on its habitat collected by Pavel, nothing is known about the biology of this mysterious creature.

Tish had sailed to the south coast of New Britain in the Bismarck Archipelago. It was our first venture in this large island group, and she had collected samples of a wide range of mammals. Among the most interesting of her discoveries was a specimen of the spectacular Bismarck giant rat. Around the size of a small cat, it's one of the most beautiful of rodents, its coat being covered with long guard hairs that shine like burnished copper.

The work done by Tish confirmed the Bismarck Archipelago as an enigma. It has one of the largest landmasses in the Pacific, and lies adjacent to New Guinea, yet it possesses few endemic species of rats and bats, and no endemic species of marsupials. Perhaps not enough time has passed since the islands were formed for new species to arise. This conjecture is supported by the fact that some of the islands in the archipelago have emerged from the sea recently in geological terms—less than a million years ago. But there's also evidence that the Bismarcks once lay far further from New Guinea than they do today, making colonisation from the mainland more difficult. We were determined to learn more. Clearly more work in the region was warranted, and the opportunity to do it would soon arise.

2

BISMARCK'S ISLES

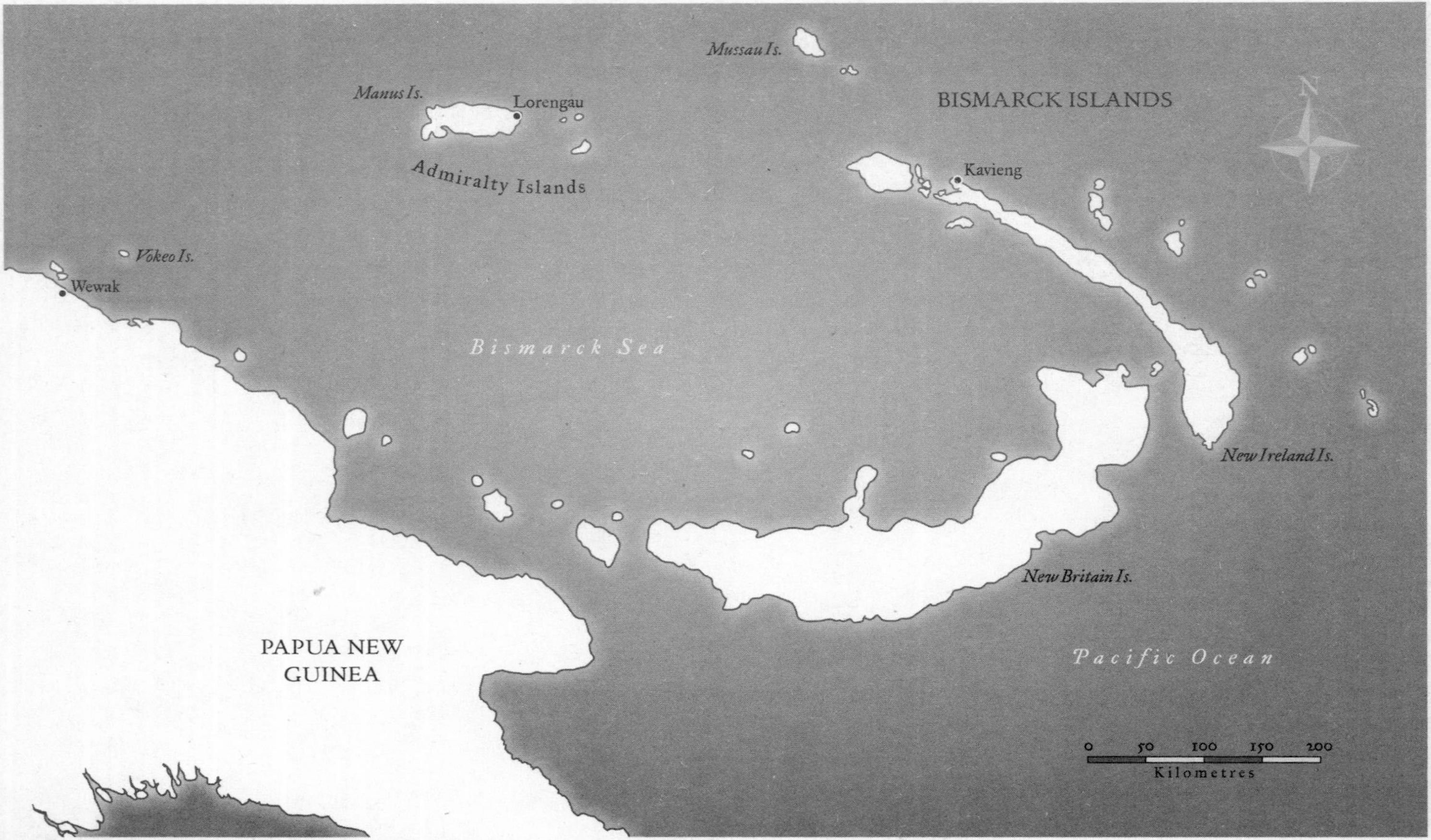

Mussau Is.
Manus Is.
Lorengau
BISMARCK ISLANDS
Kavieng
Admiralty Islands
Vokeo Is.
Wewak
Bismarck Sea
New Ireland Is.
New Britain Is.
PAPUA NEW GUINEA
Pacific Ocean
0 50 100 150 200
Kilometres
N

The human history of the Bismarck Archipelago stretches back 33,000 years to when Melanesian seafarers colonised the islands. Then, around 3500 years ago, the ancestors of the Polynesians passed through, settling on coral islets and other places that suited their ecology and technology. Both old and new settlers thrived, and today the indigenous cultures of the Bismarck Archipelago reflect both groups. The Dutch seafarers Jacob le Maire and Willem Schouten were the first Europeans to sight the islands, in 1616. Although visited by many Europeans subsequently, including William Dampier, the Bismarcks remained uncolonised until towards the end of the nineteenth century.

The earliest European attempts at colonisation were haphazard, if not insane. During the 1870s and 1880s, the Marquis de Rays, a French aristocrat, sent four vessels filled with colonists to New Ireland. The marquis, who had never been to Melanesia, was a truly dangerous eccentric and fabulist who was inspired by the journals of various explorers. In 1877 he proclaimed himself Charles, King of New France, an imaginary empire encompassing all of the then uncolonised islands of the Pacific. He claimed to want to further the interests of the Roman Catholic Church, but at the same time he swindled seven million francs from his supporters.

De Rays promoted the virtues of his imaginary realm in the pages of a self-published journal, as well as in posters and broadsheets. He claimed to have founded a thriving town called Port Breton, supposedly surrounded by fertile farmlands, in New Ireland. It was, he announced, the capital of his great empire. The scheme was convincing. Would-be immigrants boarded vessels loaded with decorative ornaments for Port Breton's imaginary Town Hall and other civic buildings, having paid large sums for the privilege. Hundreds of ordinary French, Italians, Germans and other Europeans were duped. And the results were invariably the same—death, desertion and disillusionment. But still the gullible rolled up.

The fate of the marquis' fourth expedition is perhaps the most infamous. In 1880, five hundred and seventy colonists, mostly from Germany, Italy and France, sailed for New Ireland. They believed they were headed for the bustling capital of the Empire of New France, but instead they found themselves dropped off in a godforsaken hole, a part of the island that was so rainy even the New Irelanders avoided it. Surrounded by dense jungle, the settlers unpacked their boxes of supplies on the beach, only to discover hundreds of ornate dog collars and other fripperies, and few building or agricultural tools. Within two months a hundred of the would-be settlers had succumbed to malaria, other diseases and malnutrition, and the rest had either sailed off in search of rescue or were desperately awaiting salvation. Eventually, the disillusioned survivors settled in Sydney and New Caledonia.

Such madness ended in 1884 when Germany declared a protectorate over the Bismarck Archipelago, Bougainville and northern New Guinea. It was their name, Bismarck, that stuck, and they stayed until the outbreak of World War I in 1914 when Australian troops captured the German settlements. In 1919 the region was transferred to Australian colonial control under the Treaty of Versailles, where it remained until Papua New Guinea became an independent nation in 1975. It's this layered and varied human history that gives the Bismarcks their distinctive contemporary culture.

On returning to Australia from the islands of southeast Papua, I learned that Peter White, professor in archaeology from Sydney University, was undertaking an excavation in a cave on New Ireland. He needed someone to help identify thousands of bones—the remains of ancient meals—from the deposit, and had asked me to assist. He was due to return to New Ireland in mid-1988, and thought it would be a good idea if I joined him. This was an opportunity too good to miss, for I stood the chance not only of learning about the local fauna, but of understanding, through the bones in the deposit, how the island's mammals had changed over time. So, in June 1988, I found myself on my way north again, this time by air, heading for the northernmost islands of Papua New Guinea.

CHAPTER 6

Manus

I've often envied the ornithologists I've worked with. I'm usually up at the crack of dawn checking mist-nets and rat-traps that as often as not hold nothing. They, in contrast, tend to lie in their sleeping-bags listening to birdsong. By the time I return, wet and hungry and having logged just a handful of species, they've often recorded the songs of a third of all the bird species known from the island. The compensation, of course, is that, because they're much harder to survey, there are many more undiscovered mammals than birds.

Because it is so labour-intensive and requires a great diversity of equipment, surveying the mammals of a large island in a few weeks is far beyond the capacity of one person. So as I contemplated working on the very large islands lying to the north of New Guinea, I again enlisted the help of Lester Seri and Tish Ennis. They were now experienced field mammalogists, and together we had a chance of making a successful survey. But with archaeological work, as well as a mammal survey to complete, all three of us would be kept very busy.

While planning our flight I discovered that the air service from Port Moresby to Kavieng on New Ireland was a 'milk run' which stopped off at the island of Manus, in the Admiralty Group. Breaking our journey there would not add to transport costs and, because I'd long wanted to visit Manus to study a large and distinctive kind of flying fox that lived there, I decided to spend a week surveying Manus prior to meeting with Peter White on New Ireland.

Another inducement for visiting Manus was that the head of the Papua New Guinea Department of Environment and Conservation, Karol Kisokau, who was a good friend, came from there. He'd told me lots about the place, and often urged me to visit. Sadly, we arrived in Manus as one of Karol's relatives passed away. Busy with funeral arrangements, Karol was unable to join us in the field, but he opened many doors for us, making our stay on the island much more enjoyable and social than it otherwise would have been.

Manus had been seen and charted by the Dutch explorer Willem Schouten in 1616. Because its people are brave mariners who have sailed in their ocean-going canoes as far as mainland New Guinea and, according to local legend, even Singapore, Manus has also long been connected to a wider world. And recent history has not passed it by either. During World War II Manus was a vital stepping stone in General MacArthur's advance against the Japanese, and ever since it has hosted an important naval base—Lombrum in the east of the island. Today, most Australians probably know the place as the possible location of a detention centre for asylum seekers. When we visited, Manus had a well-developed economy and education system, and had produced many of Papua New Guinea's leading citizens.

Despite its history of human connectedness to the wider world, Manus is geographically isolated, lying hundreds of kilometres west of the Bismarck Archipelago and north of New Guinea. Yet, paradoxically it has a rich fauna, including several endemic bird and mammal species. This is doubtless due in part to its large size, its rich soils and its age. But one further factor has helped to enrich it: the island lies in the path of a watery superhighway. The Sepik River, which is one of the world's largest, disgorges into the sea south of Manus. When it floods it carries huge flotillas of debris, to which clings an abundance of living things all of which are swept far out to sea.

Years after my work on Manus I saw this superhighway in action. I wanted to reach Vokeo Island just offshore from Wewak on New Guinea's north coast. It's a tiny volcanic landmass, lacking an airstrip and indeed any regular transport connections to the mainland, so I had to hire a power boat to make the journey. As we travelled towards the distant speck on the horizon, the water remained muddy and fresh far out at sea. When almost out of sight of the New Guinean mainland enormous rafts of vegetation—some carrying large trees, upright and still growing—drifted majestically past us. Such floating islands are more than capable of carrying a castaway possum or rat to a distant land, and were doubtless the principal means used by non-flying mammals to colonise far-flung Manus.

When we touched down on Manus, we were met at the airport by some of Karol's friends, who were keen to show us a good time. Indeed the expedition took on something of a festive air as we were whisked into Lorengau, the provincial capital, to check in at a motel, and lay plans for supper. As we had arrived rather late, and needed to shop in Lorengau the following morning, we decided to relax and accept the local hospitality. Dusk saw us reclining on a tropical beach, with an SP Brownie in hand, and an impromptu meal of local seafood laid out under a coconut palm.

As we talked about the island, it became obvious that the war in the Pacific had had a tremendous and enduring impact. On many islands the arrival of 'cargo' in the form of jeeps, weapons and other war materiel had sparked 'cargo cults' whose adherents believed that, if they just prayed hard enough, and built enough replica radio towers out of wood and bamboo, the cargo would return. One group in Vanuatu even collected funds to purchase the President of the United States, believing him to be the ultimate source of the cargo. But on Manus the 'cargo' was very real and an ongoing source of wealth, for at the end of the war the Americans, unable to take the war materiel home, had wrapped it in waxed canvas tarpaulins and buried it. Just the week before our arrival, one man said, a tremendous trove of corrugated roofing iron had been unearthed. It was far superior, he told us, to the junk you could buy in a store these days and was worth a fortune.

As darkness fell we were invited to a local disco. I am the world's worst and most reluctant dancer, but Lester and Tish were keen, so I went along in order not to spoil the party. The hall we were taken to reminded me of an Australian country RSL club, *à la* 1960s. There was a great band playing rock songs of its own invention, and the place was packed with locals. Tish was particularly keen to dance, and one man, noticing my reluctance to partner her, invited her onto the dance floor. The trouble was, he was not much more than a metre tall, leaving Tish with precious little to hold on to. She completed one dance with the fellow and

was returning to her bench with some relief, when she felt a tug
on her dress. The invitation to dance again could not be politely
refused, and the sight of Tish apparently dancing by herself with a
bemused look on her face was more than I could bear. So I made
the ultimate sacrifice, and after another few brackets we retired to
our hotel.

We had told Karol's friends that we wished to learn about
the large flying fox that is unique to the island and to trap rats in
good primary forest. They responded that there was little hope of
seeing the flying fox, for the population had been almost completely
destroyed by a mysterious disease that broke out just a few years
prior to our visit. Although we travelled widely we never saw a
single one, which surprised and concerned me, for large flying foxes
like to roost in large colonies and are very obvious wherever they
occur. We later learned that the species had not become extinct,
but that it would take many years for it to recover. Charting such
epidemics is of course important to conservation, as they can threat-
en island species and may be spread by human activity.

We were more fortunate with the rats. Primary forest still
abounded on Manus, and Karol's friends suggested that the
best location to base ourselves was the Department of Primary
Industry's research station at Polomou, right in the centre of the
island. Arriving there, we discovered that Polomou was a raw
frontier settlement, hacked from majestic primary forest. The trees
were pretty much the tallest and straightest I'd seen in Melanesia,
their splendour doubtless owing not a little to the deep, red soil of
the region. We had travelled by PMV—a kind of minibus service
that is the backbone of the nation's public transport system—and
were allocated a room by the staff. This allowed us to set to work
quickly, erecting mist-nets and traps and preparing ourselves for a

night's spotlighting. Lester, however, seemed somewhat reluctant to come spotlighting with me, and finally confessed that the DPI staff had invited him to a feast to farewell a local man who was leaving the island to work elsewhere. The main attraction, one of them had intimated, was to be a night out hunting for, as they put it, '*kapul e' nogat gras*'. I puzzled over the term for a moment before realising that many local girls would be present, and that the 'possums without fur' were likely to be a lot more fun than chasing rats and cuscus in the bush.

I was content to spotlight alone that night, and chose to begin work in gardens and secondary forest adjacent to the research station. Local people had told me that two species of rat inhabited the island—one was a tree-climber with red fur, the other a ground-dweller with a white tail tip. As there was no scientific record of either being found on the island, I was keen to learn what they might be. Shooting is the only practicable way to catch many arboreal rats, so I was carrying a light-gauge shotgun filled with very fine shot. Soon after setting out I came across a cocoa plantation—a favourite habitat for rats. Many of the cocoa pods had been chewed by rodents, so I stood in the dark and listened. It was a fabulous night, moonless, warm and humid, with a hum of insects all around. After a while I detected a rustle, and switching on my torch saw a large rodent with red fur in a nearby cocoa bush. It was climbing rapidly towards a tall tree when I pulled the trigger, trying to stun it so that I had a chance of grabbing it uninjured.

For a second I was uncertain of my success, but then I saw, lying on the ground at my feet, a rat the likes of which I'd never encountered before. Unfortunately it was dead and, on examining it, I saw that it resembled a tree-climbing species of rat known as *Melomys rufescens* (the red mouse of Melanesia) that abounds on the mainland—but this creature was a giant, half as big again as any *Melomys rufescens* I'd ever seen. It's an evolutionary trend common on islands that small mammals tend to become larger over time, while large ones dwarf. Back in the laboratory in Sydney, I was able to demonstrate that this rat was an undescribed species. Its nearest relative was indeed *Melomys rufescens*, and its ancestors must have become isolated on Manus a million or so years ago, having been swept there by the ancestral Sepik River. I named it *Melomys matambuai*, Matambuai being Karol Kisokau's middle name.

There was another creature on Manus I was anxious to learn about—a spotted cuscus that is endemic to the Admiralty Islands. It's only half the size of the spotted cuscuses of the mainland and has a unique, brilliantly coloured coat. The females are rusty red and black, and the males black-spotted on a white background. Although the creatures turned out to be abundant near Polomou, I had no need to hunt one. They were widely traded as a food item. Everywhere I looked in markets they hung, live, in small, tightly woven cane baskets—the equivalent of Manus chicken, and costing about the same as a roast chicken in a supermarket in Australia.

The Manus spotted cuscus's origins are mysterious. It is so distinctive in colour and body form that it must have been evolving separately from other cuscuses for a very long time, yet archaeological excavations on Manus have failed to find its remains in layers more than a few thousand years old. Perhaps it had been brought to Manus from another island, but which one exactly is yet to be discovered. Another possibility is that it is some kind of hybrid.

Because it's difficult to get to many islands I was happy to collect samples of birds or reptiles if requested by my colleagues at the museum. A number of bird species are unique to Manus, and

the curator of birds at the Australian Museum was particularly
keen to get DNA samples of the Manus friarbird. I was delighted
one afternoon to see one fly into a mist-net we had set up within
sight of the research station. I got up to search for a phial with
which to take a blood sample, but Lester beat me to the bird. He
shot out of his chair like a rocket, rushed to the net and released the
creature before I could sample it. The species was, he said, regarded
as very special by the local people, and we should not interfere
with it. Lester is a university-educated biologist, but he is also a
Melanesian. He has his own totem bird—the Papuan crow—which
he refuses to harm in any way. His guidance in such matters was
invaluable, helping me to avoid many difficult social situations.

Our stopover on Manus was too brief for a full survey, so I was
delighted that it had yielded a new species: the large, red, tree-
climbing rat, *Melomys matambuai*. There wasn't time, however, to
track down the ground-dwelling rat species known to the local
people for its white-tipped tail. At first I thought it might have been
the water rat, which is found on many islands. But in this I was
proved wrong: soon after my visit, jawbones, possibly belonging
to this species, turned up in five-thousand-year-old archaeological layers on the island. They tell of the existence of a formidable,
hitherto unknown rat with a powerful bite.

There was no way to know whether it was the species the
islanders talked about inhabiting the forests today. But then a
student from an American university, who was visiting Manus in
the 1990s, collected a skeleton of a recently dead rat whose jaws
matched those in the archaeological deposits. To this day the species
remains undescribed, and no biologist knows what the creature
looks like in the flesh. That such an impressive rat, which should be
relatively easy to find, remains unknown to science and unnamed

in the early twenty-first century illustrates how much we have still to learn about the mammals of the Pacific Islands.

Surveying for large, ground-dwelling rats like the mysterious species of Manus often necessitates the use of conventional rat-traps. They're lighter and easier to carry than the box traps I preferred that trap creatures alive. Conventional rat-traps, in contrast, kill any creature that enters them. Their other disadvantage is that they quickly disappeared from our trap lines, the local people finding them highly useful to control rats in their homes. Despite this we usually carried a few hundred, giving away any that we had with us at the end of a trip.

Because we used so many traps I had wanted to buy them in bulk from the manufacturer. The only local maker was the Supreme Rat Trap Company in Mascot, a Sydney suburb. I decided to make a visit, and was surprised by what I saw. The premises was a Depression-Era tin shed near the airport, and as I entered through a creaky door, it looked as if nothing had changed in sixty years. A glass case served as a counter, in which were displayed hundreds of rat-traps of various makes and models, and a sign saying, usefully, 'rat-trap museum'. Clearly, rat-traps, and the inventiveness behind them, were taken seriously here.

Behind the counter sat a man who bore a faint resemblance to the British comedian Benny Hill. 'How can I help you,' he said softly. As I explained what I wanted, I noticed that most of the shed was taken up with an extraordinary Heath Robinson type of machine made up of bicycle wheels, belts and seemingly endless coils of wire. In one end went wood, flat metal and wire, and as it clacked away from the other end emerged rat-traps. A counter above recorded the number produced. As I watched it turned over, with a precise click, from 23,735,491 to 23,735,492.

Astonishment must have showed on my face, for the man explained with some pride that the machine had been running since 1931, and had supported three generations of his family. It was built by his father, who took literally the saying that if you could invent a better rat-trap you'd make a fortune.

As he showed me around I noticed an old man sitting in front of an electric heater at the back of the shed, a blanket over his knees. His eyes were fixed on the machine's counter, and as he watched it record the total number of rat-traps made, a young girl brought him a cup of tea. 'That's my father, the inventor,' the man said, 'and my daughter.'

The Supreme Rat Trap Company is long gone, and the area where it was located has been redeveloped. It's hard to know, sometimes, what's changing faster: my own culture, or the cultures and fauna of the Pacific Islands.

CHAPTER 7

New Ireland, and Travelling in Time

After bidding farewell to our new friends on Manus, it was only a brief flight to Kavieng, the capital of New Ireland in the Bismarck Archipelago. At around 7000 square kilometres, New Ireland is a substantial landmass, but it is nonetheless dwarfed by its southern neighbour, New Britain—which, at 35,000 square kilometres, is one of the largest islands in the Pacific. Prior to our work, the mammal fauna of these islands was believed to be similar, for many old museum specimens were simply labelled 'Bismarck Archipelago' and could have come from either place. But geological studies of the islands revealed that they have had rather different geological histories. They were never joined by a land bridge during the ice ages, which should mean at least some difference in fauna. Tish had previously sampled giant rats and bandicoots of New Britain. Could they be found on New Ireland? And did New Ireland shelter unique species not found on New Britain?

At about 360 kilometres long, New Ireland is shaped somewhat

like a club. Its end terminates in the wild mountains of the Hans
Meyer Range of the southeast. Near the handle's tip, at the opposite
end of the island, perches the sleepy town of Kavieng, the provincial capital and old headquarters of the German administration
of the region. In the town's sandy graveyard stands the imposing
tombstone of Franz Bulominski, the most vigorous of the German
administrators. He has been described as having an 'iron hand...
fiery eye, awful presence and ruthless energy.'[7] You get a sense of
the German colonists from historic photographs. Moustachioed
Teutons clad in immaculate white suits lounge under immense figs
or in crisp tropical interiors, peering at the camera through their
monocles.

Judging from their legacy they do seem to have been diligent if
stern administrators, and their ways doubtless left a deep imprint
on the people of New Ireland. In Kavieng the colonial administrators created a model colonial German town, evidence of which
remains to this day. One relic of the era is the Kavieng Club. With
its dress rules, kitted-out dusky waiters and clubbish atmosphere,
entering it felt like stepping back in time. We enjoyed more than
one gin-and-tonic there.

Leading eastward out of the town is Bulominski's master-
work—the Bulominski Highway. It winds through coconut
plantations and past seaside villages on its way southeast, connecting the scattered communities along New Ireland's northern
shores. In 1988 I found myself travelling down that highway in the
company of Lester Seri, Tish Ennis, Peter White and his student
Tom Heinsohn. Tom was a handsome young giant, and as we
slowed to pass through villages, young women would sometimes
call out 'saizo'. The cry really is a question, deriving from the
English 'size-o?', and meaning something like 'do you fit me?'.

Even for us less physically attractive types the journey had its charms. As we wound through traditional villages and past magnificent white beaches, we encountered old plantation houses—some still operating, but most derelict relics of a colonial past. The most intriguing sight was an old house perched atop a cliff, beside which was a capacious garage full of Jaguar cars of varying vintages, all slowly rusting away in the sea air. The owner seemed to have purchased a new model every few years, and then left its predecessors languishing among the coconuts, before finally abandoning his enterprise.

Our centre of operations was to be Madina village, a rather modern-looking settlement by the sea, whose chief attraction, for us at least, was Balof Cave, where Peter was carrying out his excavations. Balof was a huge, natural cathedral, lit by sunlight from above, and with a dry floor. It was located in limestone cliffs just behind the settlement, and since time immemorial has made an excellent campsite. Although the villagers no longer used it, they had taken refuge there as recently as World War II, and there were vestiges, including empty food tins, of that occupation. Just a few centimetres under the dusty cave floor, however, were clues about the lives of far earlier tenants.

To the astonishment of archaeologists, Peter had discovered that people had been using the cave for at least 30,000 years. The island those first settlers lived in was, however, very different from the New Ireland of today. Back then, Balof Cave stood on a ridge more than a hundred metres above the sea, whereas today it is almost at sea level as the sea has risen since the last ice age. As we found while examining the bones Peter had dug up, the fauna of the island was very different too. It was, however, only with patient sorting and identification back in the lab that the full story emerged.

In the lowest levels in Balof, Peter unearthed a layer of mustard-coloured clay containing no evidence of humans and just a few scattered animal bones. This had been laid down more than 33,000 years ago—before people had arrived on New Ireland, and the bones of the animals it entombed had been either washed in, or carried into the cave by predators, such as owls. Apart from bats and birds, the fauna represented in it was meagre indeed, for it seems that just two kinds of rats—one of which had never been described and was now extinct—were the only land mammals present on the island. This astonished me, for New Ireland is a vast landmass with the potential to support a much more diverse mammal fauna. It was our first indication that New Ireland might have a very different faunal history from neighbouring New Britain.

The discovery made us wonder where all of the other mammals that we encountered on the island—such as wallabies, possums and rats—had come from. The answer came from the higher levels in Balof, for those layers were composed largely of the debris discarded by prehistoric hunters who had camped and cooked in the cave. At levels dating to around 10,000 years ago the bones of the common cuscus first appeared, indicating that the species had arrived, most likely carried by humans, from nearby New Britain. This amazed us, for it represented the earliest example, anywhere in the world, of humans deliberately translocating fauna. To judge from the thousands of jaw bones present, the cuscus established itself quickly, and must have been a godsend to those stone-age hunters.

Balof also preserves a record of plant pollen, and it suggests a downside to the introduction. At around the time the cuscus became abundant, a great disturbance of the forest canopy took place, allowing understorey plants to grow, their pollen thus increasing in the sediment. Peter initially thought that this might mark the onset of agriculture, and the felling of the primary forest by people with stone axes. But I suspect that New Ireland suffered from the introduction of a possum much as New Zealand is suffering today. In New Zealand, Australia's brushtail possum, which was introduced in the nineteenth century, was the culprit, killing entire forests and letting the understorey plants grow.

Peter's excavation revealed that around 8000 years ago the New Irelanders made a second addition to their living larder—a wallaby known as the dusky pademelon. It offered a large packet of meat, and it must have come from even further afield than New Britain, most likely from the mainland of New Guinea. Anyone who has nurtured a joey will know how easy it is to tame them. But even so, keeping a wallaby alive in an open canoe on such a lengthy sea voyage would have been a considerable achievement. Sadly, wallabies have been exterminated in the Balof area by overhunting, but the oldest island residents remember that half a century earlier the wallabies abounded.

In levels dating to the last 3000 years, Peter discovered evidence that a whole swathe of species was introduced to New Ireland— including dogs, pigs, and two kinds of rats. One, the appositely named large spiny rat of New Guinea, probably drove one of New Ireland's two original rat species to extinction. All of these creatures arrived with a new people. Known as the Lapita Culture, they are most likely the ancestors of the Polynesians. They had originated in Taiwan, and after sweeping through Melanesia went on to colonise the entire Pacific.

There was one species, however, that was abundant in the forests of northern New Ireland, but was strikingly absent from the cave sediments—the spotted cuscus. In order to uncover its history, we needed to consult one of the oldest residents of Madina village.

Sanila Talevat was in his eighties when we met him. He was short, stocky and always smiling. With intensely black skin and a scarlet-stained mouth from a lifetime of chewing *buai* (betel nut), he cut a striking figure. But what you noticed first about him was a splendid pair of snowy-white sideburns. They were real mutton chops, the likes of which I'd only seen in nineteenth-century photographs of self-satisfied European gentlemen. As a child Sanila had attended a primary school run by the Germans, and perhaps he'd received his preference for facial hair straight from the Fatherland. He could certainly shout *raus,* along with a few other phrases of German, as authentically as any Teuton.

Sanila had served with distinction as premier of New Ireland before retiring to his village. As the owner of Balof Cave he had welcomed Peter and his team to the area. In his retirement he was assisting young Papua New Guineans who had run into trouble with the law. Among those we met during our stay was a murder-er, newly released from Bomana Prison, who was from a remote village in Chimbu Province. He'd been jailed for his part in a traditional revenge killing. Such killings are considered an honour rather than a crime in his society, and the youth would have had little choice about participating. Sanila had somehow met him and, seeing the good in him, had given him a new start in life. Despite being locked away for a decade in a notorious prison, under Sanila's tuition the young man was flourishing. I was deeply humbled by Sanila's sense of humanity and his disproportionately high contri-bution to the world.

When we asked Sanila about New Ireland's spotted cuscus, his answer was precise. The creatures were, he said, descended from a pair or two that had been brought to Kavieng from Mussau Island in 1929 by a New Irelander employed by the police department.

They had escaped when their cage was upset one night and had vanished into the jungle. By 1998 the population had extended its range to about ten kilometres south of Kavieng, and their distribution continues to expand.

Sanila also told us how cane toads arrived on New Ireland. They were brought by a health officer, Mr Levi Matarai, in 1938 or 1939. In Sanila's experience, the numbers of both snakes and mosquitoes declined substantially as the toads became established. Most Australasian snakes are highly sensitive to toad toxin and often don't survive attempts to eat young toads, so I wasn't surprised to learn about their decline. But I'd never heard about toads leading to a reduction in mosquitoes. Perhaps the tadpoles fed on the mosquito larvae. Whatever the case, Sanila considered the toads a blessing.

These stories brought me a great sense of the continuity of the human traditions of New Ireland. The first settlers had arrived on an island of unfulfilled potential. It might have been an enormous landmass, but it had arisen out of the sea relatively recently and was remote. Chance rafting and drifting had brought it just two kinds of land mammals—both rats of modest size. Over the millennia the islanders had introduced one species after another, until the island was a well-stocked larder for hunters. Because there were so few native species in the first place, the disturbance (except possibly to the bird fauna, whose history remains poorly known) caused by the introductions was minimal. I was learning that not all islands are equal in terms of their biodiversity or history.

In order to identify the bones Peter was excavating from Balof Cave, I needed to collect samples of the modern island fauna. The most common mammals on New Ireland today are bats, and many bat bones were being unearthed in Balof. But conditions in the shelter must have changed over the years, for we saw very few bats there during our stay. This meant that we had to seek out other caves, which might still harbour species like the ones whose bones were being excavated in Balof, in order to collect our samples. Such caves, it turned out, were widely scattered, and we had to travel as far as the Lelet Plateau, around a hundred kilometres south of Medina, to obtain a good variety of bats. This plateau, which lies at around 1000 metres elevation, is riddled with caves, and Lester and I split up to scout them out. My caves were all pretty tame—being easy of access and with few or no bats. But Lester had a very different experience. When we met up he told me that he had entered a cave that was shaped like a funnel. Once inside, he found himself on a very slippery, steep slope heading swiftly towards the centre of the earth like Alice in Wonderland. He had only stopped himself from falling further by hanging on to a passing stalagmite.

I had an unpleasant experience of my own in a prodigious cave located in the hills a few kilometres east of Madina. It contained so many bats that a man could go deaf with the noise of their cries and flapping about, and the stench was awful. I had set out with Sanila's adopted son. He led me to a huge cavity in an abrupt limestone face. From it, a tunnel led to a steamy chamber the size of a cathedral and there, just as Sanila had described, roosted an immense colony of bats. There must have been tens of thousands of them. It seemed they were all common bare-backed fruit bats, adequate samples of which we had already collected. But in the distant torchlight on the rear wall of the cave, I could see a few smaller ones fluttering. The trouble was that in order to get to them I would have to pass through the fruit bat colony, whose screeching and flapping made the place insufferable enough. But it was the ten-metre-high pile of bat faeces that constituted their toilet that had me thinking twice about going further. The cauldron of filth had overflowed into a pool on the cave floor, and the stench was about all I could bear. Even from a distance I could see that the pile was heaving and writhing with a multitudinous army of maggots, beetles and other shit-loving creatures working across its surface.

As sick at heart as I felt at the sight, my choices were limited: having come all the way from Australia there was no way I could just turn around and leave without knowing what those smaller bats were. So it was either Bat Shit Mountain or Bat Piss Lagoon, and I chose the mountain. I tentatively edged into it and a foul, living broth slopped over the tops of my boots. The hard legs of beetles scratched at my toes, and maggots squirmed at my ankles, as my footwear filled with the shit. Higher and higher up my legs the muck crept, and as it rose above my knees I debated turning back. Then a slight firmness underfoot signalled that the worst was over. I was mistaken, however, and soon the vile, squirming mass reached my thighs and an acrid, ammonia-like stench began to suffocate me. Yet I was still only halfway to the cave's rear wall.

Then, with dawning horror, I realised that I had not escaped Bat Piss Lagoon after all. An arm of it swept behind Bat Shit Mountain, cutting me off from the cave's rear wall, and the small bats. And then came the deep bit. Plunging in to the waist-deep bat piss I wondered just how deep the lagoon might be. Relief came as my feet found the squishy, relative solidity of a rising shelf of bat night soil, and within seconds I reached the far side, without having had to undergo full immersion.

The small bats turned out to be a common species of insect eater. One that I had not seen previously on New Ireland, but it was hardly an earth-shattering discovery. I sloshed back, a little sullen-faced. Sanila's son must have sensed my mood, for he told me that there was another cave nearby that might be worth a look. So, still covered in stink, I went to investigate. This one, thank heavens, was dry, but contained only a few small brown sheathtail bats, of which we caught three. Intriguingly, I could not immediately identify them. Back in the laboratory we discovered that they were a new species. Remembering Lester's many heroic efforts in the field, and particularly his work on Alcester and the Lelet Plateau, I named them *Emballonura serii*—Seri's sheathtail bat..

Sanila Talevat was just about the last person alive who remembered the traditional names for the animals of New Ireland. Each morning I would bring him the animals we had caught in nets and traps the night before, and he would gently pick them up, one by one as if they were the most precious things in the world, and announce their names. Some were so beautiful that they sounded like fairytales. The bat known in English by the cumbersome name of the Bismarck bare-backed fruit bat, he named *Amanda Yei Laras*, meaning the bat whose fur has been touched by the sea. The creature does indeed have a peculiar, deep green tinge to the fur, as if during its nocturnal wanderings it enters the ocean and somehow carries its colour back with it.

There is another, much smaller bat on New Ireland—about the size of a starling—whose wings are spectacularly black-, orange- and pink-spotted. It is known in English as the Bismarck blossom bat, but Sanila called it *Amanda arehwak*, meaning poison bat. Sanila assured me that if ever one is heard calling at night, then it is certain there's a sorcerer about. Hearing the tale I could imagine Sanila as a kid, listening in the darkness for the call that announced an evil act—perhaps a spell to make somebody ill or to blight a crop. I was disappointed that I never heard one in the forest. In fact its call has never been recorded by a western scientist.

One morning we found a bat entangled in our nets, the likes of which I'd never seen. It was a fruit-eating species, with a body about the size of an eight-week-old kitten, and it was so subtly beautiful that I wanted to dash immediately to Sanila to learn its name. Its face bore indistinct stripes running along the snout, but its most striking features were its wings. They bore a brownish, vein-like pattern that covered a wonderfully translucent wing membrane. *Amanda ila wana aflas* Sanila pronounced slowly and deliberately, as he raised it in his hand. He had not seen one for a long time, he said, and as he looked at it he seemed to drift back in time, perhaps to when he was an adventurous young boy on the hunt investigating thickets and tangles without a care for danger. To him it was the bat with wings like the leaf of the *aflas* tree. I never discovered what the *aflas* tree was, but I suspect it was a kind of banana or near relative, for wilting banana leaves can take on the colour and venation we could see in the wings of this bat.

Suspecting that this beautiful creature might be something special I made it a priority for study. It was a kind of fruit bat new to science. Because I'm a cautious taxonomist I named it as a subspecies rather than a full species. *Pteropus capistatus ennisae* was named after Tish Ennis, whose work had done so much to ensure the success of our expeditions. Today, following further study, it's known as *Pteropus ennisae*—a distinct species in its own right. Ennis's flying fox and Seri's sheathtail bat are the only living mammals unique to New Ireland. It was a privilege to be able to name them in honour of my fellow expeditioners.

Ennis's flying fox and its relative the New Britain flying fox (*Pteropus capistratus*) are real enigmas. Known from just a handful of museum specimens, they are among the most striking of bats, having striped faces and brightly marked wings. Their nearest relatives—bats that also have striped faces—are found on the Moluccan islands about a thousand kilometres to the west of the Bismarck Archipelago. How did these creatures come to inhabit islands lying to the west and northeast of New Guinea, but nowhere in-between? It's a mystery only matched by their peculiar biology— males of the New Britain flying fox have been found lactating. They are, it seems, one of the very few mammals whose males feed their young with milk. With such phenomena still obscure, there is still much to be learned by biologists in the Melanesian islands.

The discovery of endemic bats on New Ireland greatly advanced our understanding of the zoogeography of the region. It was now clear that the island not only lacked the rich land-mammal fauna of New Britain, but that it possessed some bat species not found elsewhere. Because bats can fly over water, this suggested that New Ireland and New Britain were once further apart than they are today—otherwise the bats from the two islands would have mixed and the distinct species not evolved. To learn from a study of bats of the movement of islands over the vastness of geological time was more than enough reward for our trials and tribulations in the Bismarck Archipelago.

Sanila often expressed a culinary as well as an academic inter- est in the bats we caught and, in gratitude for all he had taught me, I skinned a few of the larger flying foxes we were taking for museum specimens, wrapped them in leaves and gave them to him to cook. It was a kindness not forgotten. When it came time for us to depart Madina, Sanila appeared with four roasted, store-bought chickens in hand—each beautifully wrapped in leaves, and cooked in a stone oven with native spices. They were without doubt the most delicious chickens I've ever eaten, and I was deeply touched, not least because they must have cost him a small fortune. One we ate immediately, while the remaining three flew—for one last time—from Kavieng to Port Moresby, where they made a splendid farewell dinner for the team.

Years later, when it came time to name the creatures whose bones had been excavated from Balof Cave, Peter White and I decided to name the extinct rat in Sanila's honour. *Rattus sanila*, Sanila's rat, may be long gone from New Ireland, but it was a true original, one of the first land mammal species to colonise the place, and one of New Ireland's most distinctive inhabitants.

3

THE SOLOMON ISLANDS

SOLOMON ISLANDS

Pacific Ocean

Santa Cruz Islands

Nendö Is.

Ukini Masi Is.

Makira Is.
[San Cristobal]

Sinalagu Harbour

Malaita Is.

Rennell Is.

Santa Isabel Is.

Nggela Group
[Florida Islands]

Honiara

Mt Popomanaseu

Mt Makarakombagu

Guadalcanal Is.

Tulagi Is.

Choiseul Is.

Vangunu Is.

New Georgia Is.

Kolombangara Is.

Bougainville Is.

Buka Is.

Shortland Islands

Solomon Sea

N

0   50   100   150   200
Kilometres

During the early days of our survey we'd concentrated on Papua New Guinea. Yet there was another newly independent Pacific Island state whose biodiversity was both extensive and poorly explored—the Solomon Islands. It consists of nearly a thousand islands which vary greatly in size, elevation, flora and fauna. Since the nineteenth century the Australian Museum had played a leading role in the biological exploration of the Solomons, and its collections held the lion's share of the world's reference specimens. Some had been obtained by naturalists working in extraordinary circumstances—while navy frigates, on punitive expeditions to avenge the killing of Europeans, shelled headhunting villagers. Some of these early specimens were unique; nothing like them has been collected since.

By the 1980s the need for more work in the Solomon Islands was becoming urgent. Its virgin rainforests were vanishing as massive, largely unregulated logging operations commenced on island after island. Mining was also opening up once-remote regions with roads and camps, allowing feral rats and cats to spread and devastate native species. In the face of these multiple threats to its biodiversity, the government of the Solomon Islands, which had achieved self-rule only as recently as 1976, could offer almost no protection. I feared that species might disappear before their existence was even known.

Working in the Solomon Islands was challenging. Its colonial history had been less than kind, and the infrastructure and resources left for the newly independent nation were woefully inadequate. Moreover, its history of European contact had been bloody and violent. The Solomons were discovered and named by Europeans long before Australia was. In 1568, at the invitation of his uncle the Viceroy of Peru, seventeen-year-old Spaniard Alvaro de Mendaña left Spain for Lima in South America. The Incas had been conquered just twenty-five years earlier, and Mendaña was surely intoxicated by tales of glory and treasure told by the surviving conquistadores.

Mendaña was only twenty-six when he set out west from Peru's Pacific coast, determined to discover and conquer his own new world. As he sailed across the widest ocean on Earth he was possibly sustained by legends of the mysterious continent of *Terra Australis*—a mythical southern land said to abound with cities of gold and other fabulous wealth. His fleet consisted of just two small wooden ships crewed by one hundred and fifty men, and their voyage was perilous. Food and sanitation were rudimentary, and in the fashion of the time they would have consigned any dead sailors to the ballasted holds of the vessels—the ballast rocks passing for consecrated ground—rather than burying them at sea. Running desperately short of water, and with the crew on the verge of mutiny, Mendaña finally sighted land in the vicinity of what is now Bughotu village on the island of Santa Isabel, in the Solomon Islands.

As he named his discovery, Mendaña's mind must have filled with images of goldmines as rich and forgotten as those of the biblical king. But his reward was to be no El Dorado. Instead, he encountered island after verdant island in a sprawling, northwest to southeast oriented, double chain over a thousand kilometres long. The lands were densely populated with a dark-skinned people, who were at first friendly, but with whom inevitably the Spaniards ended up fighting. Wherever the Spanish landed they left behind corpses and outrage, fleeing to their ships with nothing to show but a few poor provisions and captives. Dispirited, and facing potential mutiny, Mendaña decided to return to Lima. He arrived home penniless and emaciated more than two years after he had set out.

Almost thirty years later, at the age of fifty-four, Mendaña was to make one last expedition across the Pacific—this time to colonise the Solomon Islands. His fleet consisted of four ships, carrying 378 settlers and 280 soldiers, and they struck the island chain—much further south than they had on the first expedition—at the island of Santa Cruz (present-day Nendö). While today these islands are politically part of the Solomon Islands, they share greater biological and cultural similarities with Vanuatu than with the rest of the Solomons.

Conflict with the natives broke out almost the moment the colonists stepped ashore, and it was compounded by a vicious civil war among the would-be settlers, who had divided into factions. Within a month Mendaña had died from malaria, and soon thereafter the survivors packed up and set sail for the Philippines, taking Mendaña's body with them so that his remains could rest in consecrated ground at Manila.

The Solomon Islands were not lacking the precious metals sought by the Spaniards. Today the nation exports gold. But unlike the Inca and Aztec, who had amassed great hoards of gold and silver, the people of the Solomon Islands had remained in the Stone Age. They had no use for the gold nuggets gleaming in their shaded rainforest streams. So for over four hundred years the gold would lie hidden, and Mendaña's dreams of riches remained unrealised.

Thirty thousand years before Mendaña's invasion, Melanesian seafarers had arrived in the island chain. They had travelled on rafts or canoes from New Guinea, or the islands of the Bismarck Archipelago. They must have been at home on the ocean, for they brought with them their families, as well as goods and foodstuffs. The human history of the Solomon Islands is thus almost as long as that of Europe. By the time Mendaña arrived, the Solomons had developed a network of interlinked languages and cultures every bit as diverse and numerous as those of Europe itself.

In the north—in what today is the autonomous Papua New Guinean region of Bougainville—the descendants of those first pioneers evolved to have the blackest skin on Earth. They performed elaborate rituals for transforming boys into men, and lived in a matrilineal society wherein descent was reckoned through the mother's line and governed the inheritance of land and hereditary status. Further south, cultural evolution took a different twist. On Guadalcanal great terraced gardens were laid out and tended, and in Mendaña's day the island supported the highest population densities of the region. On nearby Malaita, in contrast, some of the people reclaimed land from the sea rather like the Dutch have done. They built large villages and lived

Kula big man, Woodlark Island.

The *Sunbird* at anchor, Alcester Island.

Outriggers, Alcester Island.

Kula canoe, Woodlark Island.

Local woman with black gazelle-faced wallaby, and Tish,
Boulder Camp, Goodenough Island.

Woodlark cuscus, Woodlark Island.

King rat, Guadalcanal Island.

Base camp, Mt Makarakomburu, Guadalcanal Island.

Kwaio woman with
corn-cob pipe, Naufe'e,
Malaita Island.

Naufe'e, Malaita Island.

Folofo'u and young Kwaio men, Naufe'e.

Me with a bat in a mist-net, Naufe'e.

Fijian monkey-faced bat, Des Voeux Peak, Taveuni Island, Fiji.

Poncelet's giant rat, Choiseul Island.

Flowers on Mt Dzumac,
New Caledonia.

Araucaria near Bourail, New Caledonia.

on marine resources, while highly distinctive warrior cultures developed inland. Most islands in fact developed their own distinctive culture, the larger ones supporting several cultures which could differ starkly.

The Solomon Islands rose out of the sea tens of millions of years before the first human arrived. They had never been connected to any continent, yet lay close enough to New Guinea to be settled by many creatures that flew over or floated across the sea. Millions of years before the first human foot left its imprint on a Solomons beach, reptiles, birds, frogs and mammals had all survived the challenging journey and made a home there. And many of these immigrants had given rise to spectacular creatures found nowhere else. Indeed, so rich is the islands' biodiversity that scientists journeying into their dark forests and high peaks continue to discover new species. I have with my own eyes seen creatures in the dense jungle of the Solomon Islands which, to this day, remain unnamed and utterly unknown to the outside world.

The Solomons were among the last regions on Earth to be colonised by Europeans. In part this was because their inhabitants had a fearsome reputation as cannibals and headhunters, and in part because the islands seemed to produce so very little that was valued in Europe and its colonies. The exception was cheap labour for Queensland's canefields, and it was a trade in human beings with more than a passing resemblance to slavery, known as 'blackbirding', that led to colonisation.

Blackbirding operated on the very fringes of legality. Carried out by some of the toughest captains and crews in the Pacific, it often involved kidnapping young Solomon Islanders who were then forced to work as indentured labourers for a number of years. Many died during kidnapping or in escape attempts. Those that survived were locked in holds like those of slave ships, then sold to sugarcane planters who set them to work in the canefields of northern Australia. When their time was up they were supposed to be transported back to the Solomons. But all too frequently they were not dropped at their home village, thus placing them in grave danger of being killed by people who were their traditional enemies.

The twentieth century was just seven years away when, in 1893, British colonial rule was established in the Solomons. Indeed it was with considerable reluctance that the British government declared the Solomon Islands a protectorate, their principal motive being the suppression of blackbirding. Despite the good intentions, the European impact on the islands was particularly fatal. By the 1920s blackbirding, along with the introduction of guns and disease, had left some islands, such as the 200-kilometre-long Santa Isabel, all but depopulated.

The region's reputation for savagery goes back as far as Mendaña. He records being offered a child's arm and shoulder as food on Guadalcanal—his revulsion sparking yet another clash. And it was not helped by the fatal shooting by bow and arrow of Commodore Goodenough, after whom Goodenough Island is named, on Santa Cruz Island in 1875. 'Holy Joe' as he was known to his colleagues had been on a mission to stamp out blackbirding. The death of this deeply religious and moral naval officer who had the welfare of the natives at heart left a lingering suspicion in the minds of European visitors for decades. When Ellis Troughton visited Santa Cruz in 1927 to collect mammals, he reported in the Australian Museum magazine that the people of the island's west coast were 'still dangerous' and refused to come aboard European vessels.

Although Troughton did not visit the main islands of the Solomons, which lie further north, he was told by European residents that until recently a particularly horrifying form of cannibalism had been practised there. It was said that an enclave of Malaitans had established themselves on the adjacent island of Guadalcanal, and that they would raid surrounding districts for children, whom they would then:

> feed and rear…to about the age of sixteen and sell to the natives of Malaita. The District Officer intercepted a party of these doomed ones in 1906 and found them to be absolutely without training or mental reaction, apparently quite unconscious of their fate, like any beasts for slaughter, and we were glad to hear that the abominable trade was wiped out about 1914.[8]

Cannibalism may have a long history in parts of the Solomons, as it does in much of the Pacific, but the 'savagery' of the natives towards visiting Europeans was almost certainly a result of European brutality. Mendaña found the islanders to be initially friendly— until the depredations and killings by his crew soured relations. And as the practice of blackbirding illustrates, most later contacts continued to be brutal. It was in this environment that the pioneering naturalists of the Solomon Islands worked. And none was more diligent or more successful than the English collector of natural history specimens Sir Charles Morris Woodford.

CHAPTER 8

Guadalcanal: Emperor, King and Little Pig

Charles Woodford must have been a born adventurer. He was working for the colonial office in Fiji when he first travelled to the Solomon Islands in search of natural history specimens. He based himself on Mbara Island off the coast of Guadalcanal, and his 1890 classic *A Naturalist Among the Head-Hunters* gives a lively account of the dangers and difficulties he faced while working on what was then a wild and lawless frontier.[9] The great prizes for him were the majestic peaks that dominate the centre of Guadalcanal— Mount Popomanesu and Mount Makarakomburu. They soar almost two and a half kilometres into the tropical sky—twice the elevation of the tallest on any nearby islands. Such isolated, lofty and mist-wreathed islands in the sky were, Woodford reasoned, sure to be immense treasure houses of unique biological diversity. To a nineteenth-century explorer there was the possibility that anything might live there—perhaps unknown and spectacular birds of paradise, or the world's largest butterfly—and such discoveries might make a man's name a beacon in the annals of biological exploration.

Three times Woodford attempted to scale the peaks, and each time, either through the vicissitudes of the rugged terrain, misfortune, or hostile natives, he was forced back. Once, the messenger he sent to contact the mountain people went missing, presumed eaten. Another attempt—which only got as far as it did by virtue of considerable intimidation and bribery—failed. The village at the foot of the mountain that he had chosen as a staging post was wiped out when twenty of its twenty-nine inhabitants were killed by raiders within a week of Woodford's arrival. Thankfully, Woodford was out of the village, hunting in the forest, when the raid took place.

If Woodford could not conquer the peaks, he was determined to rule over the islands. John Bates Thurston, the Fiji-based high commissioner to the southwest Pacific (and Woodford's boss) had recommended that a resident deputy commissioner for the Solomons be appointed. But the Foreign Office, seeing no way to fund the position from local revenue, refused to act. Thurston was visiting Sydney when the refusal came through, and Woodford saw his chance. Desperate for the job, he cobbled together a finance package out of existing funds, then wrote to the Foreign Office that Thurston had appointed him, and that he was already on his way to take up the post! He then rushed to Sydney, where he convinced his boss (who had a rather low opinion of the ambitious young man) to sign his extraordinary letter, and by 1896 Woodford was making his way to tiny Tulagi Island in the Nggela Group, which would form his base, and eventually became the colonial capital.

Woodford held the position of deputy commissioner of the Solomon Islands until 1914, and during his time cannibalism and headhunting were greatly reduced, though Malaita and other remote districts remained troublesome. Regrettably, administrative duties curtailed his biological explorations, but fortunately for posterity the collections he made prior to his promotion survive. They'd been sent to the Natural History Museum in London, and among them are creatures never seen by a scientist before or since.

It was Woodford's remarkable adventures and collections that first drew me to the Solomons. As I read his story and examined the specimens he'd collected more than a century earlier, I became fascinated with the man and his discoveries. Indeed, this chapter is the story of three rats he collected, two upright apes (Woodford and myself), and the island he explored—Guadalcanal.

Woodford's rats came to prominence in the year 1886. Michael Oldfield Thomas, who was then the curator of mammals at the British Museum (now the Natural History Museum, London) opened a crate which had travelled half the world to reach him. Months earlier, in a grass hut by a tropical beach on a distant archipelago, Charles Woodford had placed in it his hard-won collections, the assembling of which had more than once cost lives. Now, after travelling by sail and steam, the items contained in the crate had arrived safely at the desk of the one man who could appreciate their true significance.

Oldfield Thomas was a most particular individual. Dashingly handsome in the vein of Lord Kitchener, he was the greatest taxonomist, as classifiers of mammals are known, who ever lived. In a career spanning five decades he named over 2000 living mammal species—around one third of the world's total. This great achievement was made possible only by Oldfield Thomas's extraordinary diligence and his marriage to an heiress whose fortune permitted him to employ collectors to scour the four corners of the globe in search of novelties.

The species named by Oldfield Thomas have often been denigrated by his successors, for he frequently named them on the basis of a single individual that differed from its near relatives in only the tiniest ways. Because individuals in a population can vary greatly, such 'species' were often thought of as mistaken identifications by later generations of mammalogists. But today, with the application of techniques such as DNA sequencing, we are learning that Oldfield Thomas's eye for new species was unerring: almost invariably, when the validity of his work has been questioned, his judgment has been verified.

Great men are often greatly flawed, and Oldfield Thomas's greatest deficiency appears to have been a lack of interest in almost anything beyond his work. He was unmoved by the British countryside and—surprising in a biologist—took no interest in its flora and fauna. A colleague recalled that once, when Oldfield Thomas was looking at the beauty of the night sky, all he could think to say was that he'd classify the stars if he could. Life outside his work and marriage, which remained childless, seems to have been limited to croquet tournaments played during summer holidays at various British seaside resorts.

Our only other insights into his concerns come from a thin scatter of articles he wrote for newspapers, all of which deal with practical matters. One sketches a design for ear plugs for soldiers in the firing line, while another gives directions for pedestrians wishing to avoid collisions with vehicles and bearers of advertising sandwich boards. In others he endorses starvation as a cure for influenza, and the learning of Braille by the sighted so that they could read in bed at night without disturbing their partners.[10]

Because the field of mammalogy is so enormous Oldfield

Thomas employed several associates, the most brilliant and prolific of whom was Knud Andersen, a Dane who specialised in classifying bats. Between them, Oldfield Thomas and Andersen would do the pioneering work on the mammal fauna of the Solomon Islands.

Andersen was every bit as gifted and dedicated a taxonomist as Oldfield Thomas, and his 1912 compendium on the fruit-eating bats, Megachiroptera, is still the standard work. During World War I he was researching a widely anticipated companion volume— a complete classification of the Microchiroptera, or insectivorous bats. These bats account for almost a quarter of all living mammal species, so this was a mammoth task. Judging from his many preliminary publications in various scientific journals he had almost completed it by the end of 1918 when he vanished from the face of the Earth. The only clue to what had happened was contained in his last publication, and it was irritatingly uninformative. The article dealt with the classification of the leaf-nosed and false vampire bats. It was submitted to the journal by Oldfield Thomas because, he said, 'Dr Knud Andersen…expects to be absent from his scientific work for some time.'[11]

Rumours swirled around the museum in the years following Andersen's disappearance. Some opined that he had committed suicide in despair at leaving the manuscript of his *magnum opus* on the insectivorous bats on the train while on his way to work. Others said that he had been a spy for the Hun during the war and, fearing discovery, had fled the country. Whatever the cause, it was universally agreed that his disappearance was a scientific tragedy. To this day nobody has attempted what he so nearly accomplished, and so the world still lacks a volume dedicated solely to an comprehensive classification of the Microchiroptera.

According to his colleagues Oldfield Thomas was something of a hypochondriac—a 'perpetual valetudinarian' in the language of the day, whose hard-to-pin-down symptoms included heart palpitations and stress. He became obsessed with the effects of diet and daily massage, and he would retire to a darkened room after lunch each day for an hour-long nap. As with so many museum curators, when he retired from work in 1923 he carried on as if nothing had changed, turning up at his office punctually each day. Then, in 1928, tragedy struck. Oldfield Thomas's wife died, and after some months the grand old taxonomist seemed unable to carry on. Having long been a member of the euthanasia society he was well prepared. Evidently he had decided to die as he had lived: he shot himself with his handgun while sitting at his museum desk.

As can happen in museums, the contents of his desk remained unexamined for over thirty years. But then, in the late 1960s, John Edwards Hill was appointed curator of mammals. Like Andersen, Hill was a bat expert and, curious to see if he could discover anything about the fate of his illustrious predecessor, he turned the key in Oldfield Thomas's long-untouched desk drawers. When I met Hill in the 1980s he told me that he was astonished to discover therein a letter, which bore directly on Andersen's mysterious disappearance.

It's not surprising to a museum worker that Oldfield Thomas's desk could lie undisturbed for three decades. Museums are our best effort at stopping time. Everything in them, from a billion-year-old fossil to the skin of a rat that lived a century ago, survives in its current state only by the ongoing care of generation after generation of curators. But benign neglect plays its part too. Museum storerooms are often stacked to the roof with boxes nobody has peered into within living memory—and as long as temperature, humidity and insects have been controlled, their contents will not have deteriorated.

A story circulates among museum curators that a French museum once received a shipment of specimens from the Amazon. They'd been collected by a biologist who had vanished into the wilderness, his fate unknown. Because there was no specialist in his area of research, nobody bothered to unpack the crates containing the collection, so they sat, untouched, for a century in a storeroom. Then one day a curious young curator arrived. He opened the crates and in them discovered the smoked corpse of the collector. He'd died in the field, and his faithful Indian assistants had decided to send him home to his family along with the specimens he'd given his life for.

The letter John Edwards Hill found in the desk was in Andersen's handwriting and addressed to Oldfield Thomas. In it the Dane confessed that his life was a mess. He had married a dipsomaniac whose love affair with the gin bottle had become unbearable and, while on a visit to Budapest to study bats preserved in the natural history museum there, he'd fallen in love with an exotic dancer. He left his gin-soaked wife only to discover that while the dancer was happy to enjoy his company on a Monday, Wednesday and Friday, she could not elope with him as he had hoped because she was seeing a German count on the other days of the week. Broken-hearted, the greatest chiropteran expert who ever lived vanished. I wonder whether he leapt, despairing, into the Danube, or retreated to some secluded European village to make a garden—and never think of bats again. Unfortunately, the trail goes cold at the letter and we may never know.

Oldfield Thomas must have realised the moment he opened

the crate that Woodford's collection was a treasure trove of biological discovery. One of the most striking novelties in the consignment was a black bat the size of a cat, whose face reminded him of that of the Arctic fox. He named this great black bat of Guadalcanal *Pteralopex*, meaning winged Arctic fox in Greek (*Alopex* being the scientific name for the Arctic fox). Today we know it and its kind as monkey-faced bats. Among the most distinctive of all the creatures found in the Solomons, the monkey-faced bats are to play a key role in this story.

But even this amazing discovery cannot justly be called the pride of Woodford's collection. That honour surely goes to three kinds of large-to-gigantic rats that the explorer encountered on Guadalcanal, all of which were new to science. To them Oldfield Thomas gave classical names which translate as the emperor (the largest), king (the middling one) and little pig (the smallest).

It was these creatures that saw me, one chilly winter morning in the 1980s, standing in the London suburb of South Kensington before the great temple to nature that is the Natural History Museum. Opened in 1881, it is a grand palace dedicated to the living world. It is decorated with columns and tiles bearing images of innumerable living and long-extinct things. For a young scientist from 'the colonies', as Australia was still disparagingly referred to in London, it was somewhat intimidating going to the front desk and asking to see the curator of mammals. The post was then held by John Edwards Hill, and his reputation as a researcher of bats was legendary. I expected a distant and condescending reception, but instead was greeted warmly, and told that the collection was at my disposal.

It's hard to convey the emotions I felt on entering the great halls where the museum's treasures are stored. They are off-limits to the public, and few researchers had made the pilgrimage to examine its island rarities. Opening one large steel cabinet after another, I brought to light species of the most extraordinary kinds. Here lay the only remaining example of the giant rat of the Caribbean island of St Lucia. There, a row of skins of long-extinct pig-footed bandicoots from the deserts of Australia. And finally I located the skins of Woodford's giant rats. Giddy with excitement as I opened the drawer containing them, I hardly knew where to start my examination. I set out my measuring instruments and notebook and got going, for time is extremely precious in such a place.

Lost in my measuring and recording, I felt a hand on my shoulder. I hadn't noticed the time. It was John Hill inviting me to afternoon tea. Before leaving Australia, I'd been warned by colleagues about this ritual event. It was considered a singular honour to attend it, and here I was being invited on my first day. But how difficult it was to tear myself away from the collections! I was very glad that I did, however, for in the tearoom were assembled the heroes of my youth—curators who had discovered great chunks of the world's biodiversity—whom I could now speak to as peers, over a civilised pot of tea.

At this time I was yet to visit the Solomon Islands, and every specimen I examined was a precious source of information about this most intriguing region. The size of the emperor rat astonished me. It was a powerfully built, greyish creature about the size of a cat. Judging from its short tail, muscular forearms and the soil still lodged under the claws of a specimen preserved in alcohol, it was a ground-dweller and a capable burrower. The king rat was considerably smaller—around the size of a young rabbit. Its fur was more silvery than grey, its body proportions and feet indicating that it was most likely an adroit climber. Like the emperor, its tail was hairless, blackish and studded with tubercles. The little pig was markedly different from these two. Only a single specimen had been collected by Woodford, and very little about this portly rat with its sleek, reddish coat was revealing of its habitat. Its tail, however, was very short, making a life in the high canopy unlikely.

All three clearly belonged to a rodent genus known as *Uromys*, meaning naked-tail mouse. They're an ancient breed of murid that scientists suspect was among the first to make its way from Southeast Asia into Australia and New Guinea. They'd probably arrived by four million years ago, and had subsequently diversified into about a dozen species. But there was a profound mystery here. Guadalcanal sits in the middle of the Solomon Islands chain, and the genus *Uromys* had not been recorded on any other island in the group. Nor was it present on New Ireland, which is a logical stepping stone for any migrants to the Solomons. How had the ancestors of the emperor, king and little pig travelled from New Guinea to Guadalcanal without leaving populations on any island in between?

Several explanations were possible. Perhaps the ancestral *Uromys* had drifted directly from New Guinea to Guadalcanal, or perhaps it had colonised other islands but had become extinct on those. But it was also possible the emperor, king and little pig were not really members of the genus *Uromys* at all, but had come to resemble them through the process of convergent evolution. Convergent evolution is a common phenomenon, accounting for the similarity between the thylacine and the dog, for example, and among the rodents it is widespread. The best way to test for convergent evolution is to compare the DNA of the species in question, and that meant an expedition to Guadalcanal, for at that time it was not practicable to use DNA from museum specimens for such tests.

The hitch was that neither emperor nor little pig had been seen for over a century, and the king not for sixty years. The extinction of island species had progressed apace in recent decades, and perhaps a trip would be futile. Obstacles seemed everywhere, not the least of which was the question of where a young researcher, whose track record was sketchy at best, was to turn for funding for such a quixotic expedition. The National Geographic Society seemed my best hope, and upon returning to Australia I penned a modest proposal aimed at discovering if the emperor, king and little pig still reigned over their distant isle. To my surprise the society agreed to finance my expedition—to the princely sum of US $7000. It was my first significant research grant, and I was determined to make every one of those dollars go as far as possible.

Most of the money went into airfares and preparations for a trip to Guadalcanal. It was my first visit to the Solomons, but as grateful as I was for this grant, I soon found that there's only so much you can do with small grants and assistance from the Australian Museum Society. If I was to realise my full ambitions of surveying every major island group in the southwest Pacific and Moluccas, then an adequate, ongoing source of funding had to be found. When it did finally emerge, it came from the least likely of places.

In 1988 I had appeared on a popular television program to speak about my research on tree kangaroos in New Guinea, and a few days later I received a letter from a solicitor in Sydney, requesting a meeting. He explained that one of his clients had left a bequest to fund wildlife conservation. It was a subject he knew little about, and he wondered if I might want some of the money?

The bequest, I learned, resulted from the determination of an ordinary Australian woman to make a difference to the conservation of our region's most endangered species. Winifred Violet Scott was one of seven siblings, none of whom married. Each had lived a modest life, making prudent property purchases and investing wisely in stocks and bonds. They bequeathed their estates to their surviving brothers and sisters, until, finally, Winifred found herself the owner of seven North Shore houses and seven substantial share portfolios. In short, Miss Scott was worth a small fortune. But she did not change her frugal ways, and shortly before her death she met with her solicitor and explained that she wanted to leave all of her earthly possessions as a bequest to preserve endangered species.

The solicitor had at first tried to dissuade Miss Scott from this course, suggesting that instead she might like to consider a donation to cancer research or perhaps to underprivileged children. Miss Scott, however, was resolute in her wish, and the change to her Will was made accordingly.

When Miss Scott passed away, her solicitor discovered that the value of the bequest was even greater than he had imagined, amounting to a sum that he was at a loss to know how to disburse. He had considered donating the lot to the Stockman's Hall of Fame in Queensland, but such a gift did not fall within the parameters of the Will. As the solicitor and I spoke, it became clear that a great faunal survey of the islands of the southwest Pacific region, for the purpose of identifying its endangered species, would be something Miss Scott would have been very proud of. So, in August 1989, the Scott expeditions were born, and for five years our survey teams ranged far and wide, from the Moluccas to Fiji, making biological discoveries neither I nor anyone else had ever dreamed possible.

So successful was our expeditionary work that we were soon in need of increased assistance from the museum's taxidermy department. George Hangay had done his best to keep up with the stream of specimens requiring cleaning and stuffing, but it was a losing battle. Finding a professional taxidermist is not easy, but the backlog mounted up as the search went on. Finally, a solution dropped into George's lap in the form of a young Chinese immigrant who arrived unheralded at the museum. Alex Wang spoke almost no English, yet managed to convey that he had recently arrived from Urumqi in China's remote Xinjiang province, and was looking for work. He had no certificates or letters to testify to his past employment or skills. Instead he opened a long trench coat; pinned to the inside were the stuffed skins of dozens of small birds and mammals.

George decided that the best thing to do was to ask Alex if he'd like to volunteer at the museum for a week, so that the quality of his work could be assessed first-hand. All seemed satisfactory until Alex approached George at week's end with his hand out for payment. Alex's English was all but non-existent, and the word 'volunteer' had not been understood. But Alex's work was exemplary, so George paid him out of his own pocket, and made arrangements to employ Alex permanently in the department. With Alex's help the backlog of specimens was soon decreasing. We now had the capacity to write up our results, and a rush of scientific publications ensued.

There is only one town of any size in the Solomon Islands, the capital Honiara on the island of Guadalcanal. Woodford's old capital of Tulagi was abandoned following the Japanese invasion in 1942. When I first arrived in 1987 the nation had been independent for just a decade. Flying into Henderson Field just outside the capital I could see the majestic, cloud-wreathed summits of Makarakomburu and Popomanaseu beckoning through the aircraft windows. Was that where I might find Woodford's rats? Closer to the coast bands of dead trees revealed that the forests at lower elevation were being cleared with axe and fire. I was determined, one day, to scale the peaks that had defeated Woodford, but this was not the moment. I was too inexperienced and, despite the inauspicious forest destruction, my search for the great rats would begin closer to civilisation.

In the years that followed that first expedition, Honiara was all but destroyed by a civil war between migrants from the adjacent island of Malaita and the natives of Guadalcanal. When I did my work the war was no more than a distant rumble expressed in shouting matches and the occasional brawl. But as the years passed I watched the news, horrified, as the conflict escalated from one fought with bows and arrows, to one waged with rifles and armoured vehicles. One of the most bizarre scenes involved a formidable, locally made 'tank', which was used to flatten villages and lead assaults on enemy positions. It was crafted from a huge bulldozer that had been stolen from a goldmine and covered over with rough, welded-iron plates, every gap between them bristling with shotguns, arrows and rifles. As it ploughed through the jungle it had all the brutal, end of days craziness of a scene from a Mad Max movie.

CHAPTER 9

Probing towards the Summit

Prior to my arrival in Honiara I'd written to David Roe, an archaeologist working in the Solomon Islands who'd excavated a cave north of the capital on the Poha River. Among the bones he'd unearthed there were some jaws of the emperor rat. They looked fairly fresh, and I felt it was possible that the species still inhabited the area. So when David offered to introduce me to some people in a village near the caves, I jumped at the chance.

The village, which turned out to be almost on the outskirts of Honiara, was a rendezvous for prostitutes and their clients who arrived by taxi at all hours of the night. It was also close enough to town for the young men to get drunk and then go home and cause trouble. I hardly slept at all on my first night there for the screamed abuse of drunks, some of which was directed at me, and the comings and goings of the prostitutes and their clients. But, worst of all, the village was a long way from the forest.

My brief stay did, however, have one great benefit, for I met some older men who knew at first hand of Woodford's rats. One even claimed to have captured a *kandora mbo*, as he knew the emperor, around the time of World War II. The name means 'ground-living possum' and he described finding them in burrows, so confirming my suspicions that the species was terrestrial. He advised me to walk into the foothills where the forest was better, and search for the creatures there. I didn't need much encouragement, and I set out late that morning in the company of some local landowners. Loaded down with equipment we walked through lowland grasses and decimated re-growth. It was suffocatingly hot and I was delighted when we came to a steep valley that had retained much of its original cover of trees. Even better, here the Poha River formed a magnificent waterhole, its crystal-clear waters flowing over a sandy bottom before cascading through a rocky defile. The water was alive with freshwater prawns and tiny coloured fish. It was a perfect place to camp.

That evening, after a refreshing swim, I prepared a meal of rice and mackerel—typical fare for fieldworkers in the Solomons. As the sun set, the distant siren-like sound of the six-o'clock cicadas rose and fell with the last of the light, and slowly the creatures of the night roused themselves. Most of the mammals of Melanesia are nocturnal, so for a mammalogist this is the most exciting moment of the day. I was not prepared, however, for the experience of having a huge diadem horseshoe-bat—its wingspan the length of my forearm—land on the branch of a dead sapling just a metre in front of me as I sat in camp. Its intricate nose-leaf twitched as it sent out its ultrasonic, pulsed cry with which it reads the world. As the sound patterns bounced back to the restless creature it twisted its body this way and that, rotating its ears like radar dishes. How, I wondered, did I appear in its ultrasound world?

Perhaps its sonics permitted it to 'see' right through me to the meal of tinned mackerel and rice I'd just eaten. Or perhaps it saw my very life force—the blood flowing through my veins and the electrical discharges that powered my brain, allowing it to anticipate my reactions and thoughts. Whatever it experienced did not hold it, for soon the majestic creature set off in pursuit of beetles and other insects. Finding its food is quite a job; the sixty-gram bat has to consume its weight in insects every night just to stay alive.

As full darkness closed in, bats seemed to flit everywhere. Tiny ones brushed my face as they chased mosquitoes and great fruit bats whooshed their wings overhead as they set out for distant fruit trees. The Solomon Islands really is a world of bats, for no other mammals apart from the rats had reached these distant outposts before the arrival of humans. The fauna has of course been enriched since then—first by a cuscus, most likely carried from the Bismarck Archipelago six thousand years ago, and then, in the past three thousand years, by pigs, dogs and commensal rats. Europeans have added cats and cattle, both of which have run wild. But, even so, the bats continue to dominate. The insectivores range in size from tiny, nondescript cave-dwellers, through to a gigantic relative of the diadem horseshoe-bat I'd just seen, which is one of the largest insect-eating bats. Then there are the fruit bats, ranging from the elfin, orange-, black- and pink-spotted Solomons blossom bat through to the largest and arguably the strangest bat of them all—the awesome, giant monkey-faced bat of Bougainville. Many of these bats are found nowhere else, and their ancestors must have been among the first mammal colonists to reach the islands.

As I pondered this marvellous, flittering diversity, a huge, completely black creature rose from the canopy a hundred metres away. It was a Guadalcanal monkey-faced bat—the first monkey-face I'd ever seen in the flesh.

Because most fruit bats, logically enough, eat fruit, they have simple teeth and weak jaws. The monkey-faces, however, are different. Their ancestors must have been among the first mammals to arrive in the Solomon Islands, discovering there a terrain with no land mammals except, perhaps, for the rats. The trees in their new home were full of nuts and tough fruit—from coconuts to Pacific almonds and candlenuts. On the mainland various marsupials and rodents have evolved to crack open such nuts. The fruit bats, in contrast, sip nectar from flowers or eat soft fruit. Their molars are rudimentary as, presumably, were those of the ancestral monkey-faces. But as they began to try to crack the tough fruit and nuts, those that succeeded without breaking their teeth were favoured in the race for survival.

As a result, today's monkey-faced bats have complex molars covered in tough enamel, and enormously strong cheek muscles. There are five species in the Solomons and one in Fiji, the largest of which has such massive teeth and jaws that it can crack young coconuts. All species have short, powerful muzzles, small ears hidden in fur, and striking red or orange eyes that endow them with a rather lemur-like appearance. Hence their common name.

I feel a strong link with the monkey-faced bats, for one bears my name. It was described by Kris Helgen. He was my doctoral student and now heads the mammal department at the Smithsonian Institution. During his studies he resolved a problem that had plagued mammalogists since the days of Oldfield Thomas and Knud Andersen. Both men had described species of gigantic, black monkey-faced bats from the Solomon Islands—Oldfield Thomas's from Guadalcanal and Andersen's from Bougainville. Although the Bougainville population seemed to be very variable,

no researcher had seen enough of these rare bats to publish a definitive classification. After examining every museum specimen in the
world, Kris realised that the Bougainville animals consisted of two
distinct species, whose distribution broadly overlapped. The largest
kind—a black creature with the wingspan of a small eagle—had
never received a scientific name. Its teeth and jaw muscles are
enormously powerful, doubtless enabling it to crack the hardest
nuts in the forest. In 2005 Kris named it *Pteralopex flanneryi* in
recognition of my research in Melanesia.

No photograph of a living Flannery's monkey-faced bat exists,
and nothing definitive is known of its ecology. Even though it's
named after me, and I've worked in the Solomons, I'm yet to see
one in the wild. If I want to do so I'll need to hurry, for, like many
of the Solomons' most distinctive and ancient inhabitants, it needs
primary forest and those forests are fast disappearing. Indeed all
of the monkey-faced bats are listed by the International Union for
the Conservation of Nature as either endangered or critically
endangered.

The monkey-faces have evolved a neat trick that makes them
particularly graceful fliers. Their wings meet in the middle of their
backs, giving them a huge wingspan-to-lift ratio. This means that
even though they can weigh a kilogram they're able to fly slowly
and precisely—even backwards if the need arises—an adaptation
that may be vital as they search for food in the complex spaces
under the forest canopy.

The presence of a Guadalcanal monkey-faced bat on the Poha
River augured well, I felt, for my search for giant rats. When I set
out that night accompanied by two local lads, my legs were already
aching from the day's exertions, and I was desperately tired. To
make matters worse, the track we followed was rough, forcing us

to scramble over boulders and down slippery slopes. And all the time I had to keep my spotlight on the canopy and my shotgun on my shoulder. We had barely gone fifty metres when I noticed a reed beside the track moving in an unusual way. As I watched, a red, kitten-sized creature climbed up it. It was a rat the likes of which I'd never seen, and once it had emerged from the surrounding vegetation it sniffed myopically in our direction. Hoping that we might be able to capture it alive, I kept the spotlight steadily on it, while signalling to one of my companions to move in behind. I was both astounded and dismayed when he casually reached forwards as if to grab the creature in his hand. In an instant it vanished, leaping from the reed into the dense foliage beside the track. Looking confused, he explained that he thought it was a baby cuscus (which is also red)—a creature so slow and inoffensive that it can easily and safely be caught by hand.

In all my time in the Solomons, this would prove to be the first and last time I ever saw the mysterious reddish murid. Was it Woodford's little pig, a creature known from just a single specimen collected over a century before? I've thought deeply about it over the years, trying to recreate in my memory the details of that night. Yet even today I cannot be certain whether I'd encountered the little pig or some other species that remains unknown to science. If I were forced to bet, I'd incline to the view that it was an unknown species, for its agility in climbing that stalk sits ill with the short tail and chubby build of the little pig specimen I'd seen in the Natural History Museum.

But the experience made one thing crystal clear to me—the importance to a field biologist of obtaining a specimen. Without the physical evidence of what you've seen, you can never be sure of what kind of creature you've been observing, nor can your discovery be verified. This means, apart from anything else, that endangered species can never be assisted without a specimen. The little pig is presumed to be extinct today, but if I had been able to establish that a relict population survived in the Poha Valley, then measures to preserve it might have been put in place.

For several hours after that exhilarating moment we saw nothing but possums, roosting birds and bats. All were interesting creatures to observe, but none was what I had come for, and I was about to head home to bed when, high overhead, on a liana connecting two huge forest trees, I thought I caught a glimpse of silver. Craning my neck, I could vaguely make out a rat-like shape against the dark sky. A shotgun—even one filled with the lightest shot—is a blunt instrument for collecting mammal specimens, and I hate using one. But when I was studying arboreal rats in the 1980s it was the only means of gaining a positive identification.

The silvery gleam had almost disappeared into the foliage by the time I pulled the trigger. Because I was shooting directly above my head, the kickback from the gun almost dislocated my shoulder. But then a magnificent, silvery male rat the size of a young cat dropped to the ground at my feet. It was *Uromys rex*—the king rat—and it was stone dead. I felt a moment of regret, but when I looked into its mouth to confirm my identification, I didn't feel so bad. It was a geriatric male whose molar teeth were so worn that little but roots remained. Without teeth, even if I had never seen it, its life was most likely counted in weeks rather than months or years.

The sacrifice of that one geriatric rat would provide the key to understanding much about all of the rats of the Solomon Islands. Its tissues permitted a DNA analysis which confirmed that the king rat was related to the giant naked-tailed rats of Australia and New Guinea, even though it had separated from them aeons earlier. So, several million years ago, a naked-tailed rat must have made an astonishing journey from New Guinea to Guadalcanal. The distance involved is enormous and, a bit like Mendaña, that first voyaging rat had made landfall on an island in the middle of the archipelago. We were to learn far more than history from the sacrifice, for by studying its morphology and stomach contents we learned about its lifestyle and diet. Like so many of the ancient endemic species of the Solomons, it's dependent upon primary forest for both food and shelter. But above all the specimen was important because it constituted proof positive that the king survived and that measures to conserve it would not be undertaken in vain.

But what of the emperor? Did that largest of all the Solomons rats still survive in some remote corner of the islands? Further work in the Poha River area, as well as in other lowland regions of Guadalcanal, had turned up nothing definitive. But old hunters who knew the creature at first hand said that, although they had not seen it for decades, it probably survived in the mountains. Clearly I needed to attempt what Charles Woodford was never able to do—ascend Mount Popamanesu or Mount Makarakomburu. It was to be a far more prolonged and difficult undertaking than I ever imagined.

When I returned to the Solomon Islands in 1987 I was intent on taking up the challenge, but fate waylaid me. I'd been lucky enough to meet up with an Australian photographer, Mike McCoy, who was married to a woman from Malaita and knew the Solomons intimately. Indeed he was a bit of a celebrity among the islanders, having produced many of the iconic postcards sold in local shops. Over the years Mike and I were to have a lot of fun together. When I explained that I wanted to climb the mountains, he said that a great deal of planning would be required, and that while he put feelers out we could fill in time exploring places closer to Honiara. The first place he led me to was the Gumburota Caves behind Honiara, where there might be bats. The caves had been the scene of heavy fighting during the war, and Mike had recently found a thigh bone in the bush there. It was from a very tall person— probably an American soldier—and it had a sword cut passing almost through it. As we stood before the caves I could imagine the tall G.I. standing where we now stood while a Japanese officer rushed suicidally out of the darkness, raised samurai sword in hand. The American, who must have been alone, had probably bled to death in the bush. For half a century he must have been missing in action, his family having had no idea of what happened to him. The tragedy of a war whose wreckage I had seen everywhere in Melanesia suddenly became very real to me.

The Gumburota Caves on the Kehove River near Honiara are deep and wet, with a rivulet running right through them, and they are full of bats. One of the many shapes I saw flitting through the torchlight was that of just about the largest insect-eating bat I'd ever seen. We strung a mist-net across the cave and it quickly snared what turned out to be a Solomons giant horseshoe-bat. It's the largest insect-eating bat in all of Melanesia—and is found only on islands of the central and northern Solomons.

The horseshoe in its name refers to a fleshy growth that dominates the face. Roughly horseshoe-shaped and made up of three layers of 'leaflets' that help to capture and interpret the sound-waves that the animal uses to navigate and find its prey, it looks like a fleshy flower and it dominates the face. The wingspan of these magnificent bats approaches sixty centimetres, and they can weigh almost eighty grams. With truly formidable teeth and jaws giant horseshoe bats are capable of crunching the largest and toughest of beetles. Yet the one I drew out of the net was a gentle creature that was not inclined to bite me at all as I measured and photographed it. Later surveys revealed that it was a rare species indeed, and one that is possibly declining due to deforestation and other human impacts. Sadly, I would never again encounter as large a colony (consisting of a few dozen individuals) as I saw that day.

You never know what you are going to bump into in caves, and as I made my way back to the entrance I was surprised by the sound of something striking the cave wall beside me. Flashing my torch around I saw an enormous frog, its eyes reflecting in the beam. Known as *Discodeles guppyi*, this Solomons speciality is the size of a soup bowl and has eyes larger than marbles. The largest frog in all of Australasia, it's a striking example of island gigantism, and it provides an ample meal to villagers wherever it occurs.

On that trip I was to see other strange reptiles, including frogs that looked like brown leaves that had fallen from rainforest trees and a sinister-looking creature known as the giant crocodile skink. These weird lizards had been collected by Mike in the Shortland Islands—the very northernmost islands in the political entity that is the Solomon Islands. Black, and around twenty-five centimetres long, they were covered in knobbly, crocodile-like scales, and the skin surrounding the eyes was blood red. Their jaws looked powerful, and Mike said that they could give a nasty bite.

As interesting as they were, the crocodile skinks and the frogs and bats of the Gumburota Caves were a sideshow. It was the giant rats that I was here for, and I was determined to search those fabled mountain forests where they might still reside. Mike suggested that the easiest way in was to follow the road up to Gold Ridge, where there was a mining camp. Located at 600 metres elevation behind

Honiara, it seemed like it would be a simple matter to ascend the peak from there. But the goldmine has brought great social disruption to the mountain people, as well as a greed that, from what I heard, sometimes rivalled that of the old would-be conquistador Mendaña himself.

The track to the mine winds up deforested ridges until you reach the cool of the middle elevations where the operation is located. People from all over the mountains were beginning to congregate there, and a messy squatter camp, with all its attendant social problems, was beginning to form. We pulled the car up near some huts and tried to locate the head of the local village council. But we had hardly begun speaking before a large, overweight and dishevelled man emerged from a hut. With bloodshot eyes and bandages on his legs and arms, he looked as if he'd had a big night. Clearly unhappy with our visit, he explained that he owned all the land above the camp, then vehemently refused us permission to go any further unless we paid him $2000. This was a large sum of money for us—indeed it would bankrupt my grant—so we asked if we could just cross his land for a lesser sum. Louring and clenching his fists like he wanted to punch me on the nose, he angrily refused. In the Solomons, local landowners reign supreme, and in the face of such hostility there was nothing that we could do but return to Honiara and plan again.

On mature consideration, Mike suggested that we explore elsewhere in the Solomons for the time being, deferring the trip to the mountains until another visit. I reluctantly agreed, while dismally beginning to feel that Woodford's luck might also be my own. One good thing did seem to come out of the visit, however—in the camp we met a young man called Peter who came from a village called Valearanisi on the far side of the mountains. Known as the weather coast, parts of the region receive in excess of eight metres of rain a year, and they remain about as inaccessible as they were in Woodford's day. Peter reckoned that if we could just get to Valearanisi our problems would be over. The people were friendly, he assured us, and the place was crawling with giant rats. One old man who lived there was even known as *Hue Hue*, the local name of the Emperor rat, on account of the vast numbers he had captured.

Contemplating a visit to the weather coast, I began reading all I could about the place, and what I learned worried me. The region's reputation for lawlessness was second to none in the southwest Pacific, life there largely being controlled by warlords for whom violence was as much a way of life as it was for Guadalcanal's warriors of old. Going there to climb Mount Makarakomburu would, I feared, be a bit like stepping back to the days of headhunting and cold-blooded murders.

If the politics of the region were bad, the logistics of visiting the weather coast were no less daunting. There was no airstrip, and the weather, as its name suggests, is treacherous. Time was needed to plan properly, and I had to acquaint myself more widely with the Solomons and their people before making such a trip. Guided by Mike McCoy, I decided to visit the islands of Makira and Malaita before trying the ascent of Mount Makarakomburu. These islands are located at the eastern end of the main Solomons chain, and both were poorly explored biologically, promising rich harvests of unknown species.

CHAPTER 10

Makira's Mysterious Rat

The island of Makira, previously known as San Cristobal, is the most far-flung of the larger Solomon Islands. Lying far to the southeast of its nearest neighbours Guadalcanal and Malaita in an immensity of ocean, it's a rugged, vast and thinly populated realm of limestone, dense jungle and isolated fauna. Despite the distances involved, the ancestors of most of the creatures inhabiting the island must have arrived there from Guadalcanal or Malaita, just as I did on my first visit in 1987.

Among those lucky founders were a number of birds that would go on to populate Makira with unique and striking species. They must have arrived a million or more years ago, and over time evolution had transformed them. One distinctive endemic is Sclater's honeyeater, a large honeyeater with brown feathers, a long ivory bill and a pale eye. It has been exceptionally successful, and is commonly seen, even around villages. Other species unique to the island, however, are rarer and little known. The Makira flying fox and the Makira horseshoe-bat were each known from only a handful of museum specimens, most of which had been collected in the nineteenth century. But there was one other creature that drew me. This one was almost a complete mystery, for nobody knew which island it came from, or even what it looked like. *Solomys salamonis* is known from a single damaged skull, which had been collected over a century earlier. There were good reasons, for anybody interested in solving the mysteries of its appearance and habitat, to begin their search on Makira.

Everything we know about this enigmatic rodent we owe to a murder. In 1880 Lieutenant Bower of HMS *Sandfly* was cruising the Solomon Islands, searching for miscreants on a lawless frontier. Bower had a magnificent physique and a reputation as a first-class front-row forward in the intensely physical game of rugby. Reckless, arrogant and racist, just before his death he was seen swinging a native club vigorously above his head and shouting, 'I say, you fellows, just think of an Englishman among a crowd of these Johnnies armed with one of these.'[12]

Bower was caught, literally, with his pants down—killed while bathing on Ugi Island. His head was added to the islanders' trophy collection. HMS *Cormorant* was dispatched to avenge his death, and the Australian Museum's taxidermist Alexander Morton managed to find a bunk on the vessel. We can only imagine the conditions Morton worked under, but somehow amid the shelling and slaughter he managed to collect a giant rat and to preserve its skull and skin. It was described by an early curator, but lamentably the skin decayed and was eventually discarded. All that remains today is the skull, which is so distinctive that it could not possibly be confused with that of any other rodent.

But where, precisely, had Morton collected the rat? Was it from

Ugi Island in the Nggela Group (near Tulagi), or Uki Ni Masi (also known as Ugi) adjacent to Makira? Nobody knew, but Makira was as good a place as any to start looking, for it was far larger than the other islands and, if the creature was indeed from Uki Ni Masi, it would surely occur on Makira as well. At the time I lacked the finances to mount a large expedition—especially one premised on such slender evidence—so I decided to travel to Makira on the cheap, and alone.

It's always an interesting experience to arrive in an unknown village unannounced and unexpected. But on this occasion a stroke of good luck befell me, for sitting next to me on the plane was a man who announced himself as Makira's Minister for Sport. Before we touched down he'd asked me for a pair of Dunlop Volley sandshoes. When I said I'd post them to him, he agreed to assist me in my work.

That night I slept on the porch outside the government offices, which consisted of a few bush-material huts, beside the airstrip. The next morning, as I set about making breakfast and arranging my equipment, I was approached by a tall man dressed in an immaculate white shirt, slacks and shoes, holding a clipboard. His manner was polite, though very official. He began by asking to see my passport and collecting permit, the latter having been issued by the Department of Environment in Honiara. After carefully noting the numbers on both documents, he asked me where I had come from and where I lived. But as the conversation proceeded his line of questioning took a peculiar turn. 'What is your religion?' he asked. This is a standard question from officials in Indonesia, so I wasn't too taken aback, answering that I'd been brought up Catholic. He took care to note this, then asked very seriously if I had ever had sex with my wife. Flabbergasted, I looked up and caught a glimpse of a group of children peering from behind the offices. When they erupted in gales of laughter, I realised I'd been had. The supposed official was in fact the local village clown, and he'd taken me in hook, line and sinker.

This introduction to Makira left me wondering whether the only duty of the island's Minister for Sport was to cadge tennis shoes from visitors. He did, however, turn up later that morning and invited me to stay in Sesena Village, just a few kilometres walk from the airstrip. With the help of some young lads, I relocated, set up mist-nets and rat-traps, and began to sample the island's unique fauna.

Being close to the administration centre, I soon discovered that a British magistrate by the name of Shipley and his wife and brother lived on the island. They welcomed me into their company and over convivial lunches explained why they were there and what they knew of Makira. It may seem strange that British circuit magistrates were still being dispatched to the Solomons a decade after the islands had gained independence. But the new nation still laboured under British law, and if it was to be properly adminis-tered then British magistrates were required. In the 1980s they were still being dispatched from London to sit in judgment in villages whose inhabitants would have found the Old Bailey as alien as the dark side of the Moon.

British law sits uneasily with most Solomon Islanders. Offences that they considered most grave—such as adultery and witch-craft—are not even considered criminal under English law, while acts that they considered valiant deeds, in fact obligations—such as the killing of a man who has had sex with a female relative—are punishable by long stretches in prison. In colonial times an appearance in court must have been truly puzzling to them. As the

eminent anthropologist Roger Keesing and historian Peter Corris
write:

> An assassin in a blood feud, whose homicide was cultur-
> ally legitimate and even a duty, would find himself before a
> bewigged and unintelligible magistrate, then imprisoned in
> Tulagi [the colonial capital] for weeks or months while his
> crime of breaking an alien law he had never heard of was
> reviewed in Fiji, then led to the gallows.[13]

Even by the 1980s the court process could still prove confusing.
Over one lunch, the visiting magistrate on Makira told me of a
case he had presided over on Guadalcanal. A young woman had
brought a charge of theft against her stepfather. The magistrate
was surprised to learn, through an interpreter, that the theft had
involved milk, for fresh milk was all but unavailable locally. As the
facts unfolded it became clear that the woman had recently given
birth to a baby girl whom she was breast-feeding. She had awoken
one night to discover her stepfather, rather than her baby, content-
edly suckling at her breast! The magistrate expressed uncertainty
whether theft was the best description of the crime committed,
but nonetheless levied a stiff fine, and issued a stern warning to the
errant stepfather.

Pigs are a prime source of litigation in Melanesia, and cases
involving damage to gardens by marauding swine can become
the focus of deadly feuds which can last for generations. Mike
McCoy told me about one such trial on Malaita, which seems to
have caught the spirit of confusion at the heart of the administration of British justice in Melanesia. A villager had 'courted' his
neighbour for damage inflicted on his garden by the neighbour's
pig. The case was heard in the sleepy regional centre of Auke, in a

courtroom improvised from a local bush-materials classroom. With no air conditioning, the elderly bewigged and garbed magistrate was hot, impatient and much bothered by the flies that swarmed through the open windows. As he understood no pidgin, he was entirely reliant upon a translator.

Mr Serana, as we will call the complainant, was in his fifties, short and nuggety and dressed in only a pair of ragged shorts. He spoke no English—only the pidgin of the islands—and to him the issue was deadly serious, for, if not settled in court, it would end up erupting into rough-and-ready village justice. He began his testimony with a long description of his fine garden with its taro, sugarcane and sweet potato, and the great effort he'd gone to in fencing it to keep out pigs. Only a truly evil and determined pig, he averred, would dare breach those defences. With increasing anger he told how, while sleeping in the garden after a long day's labour, he was woken late on a moonless night by the sound of a snuffling swine. 'Thief!' he cried as he raced towards the noise. But this was a sneaky pig, rotten to the core and very experienced in stealing from other people's gardens, and it had silently slunk off, leaving only hoof-prints in the mud and a distinctly male piggy odour in the air.

At this point in what was becoming a long story, the magistrate impatiently intervened. 'But Mr Serana, did you actually see the pig?' Serana had to admit '*me no lukim*'—he had not seen it. But then he had no need to. He knew precisely who was guilty, for the size and temperament of every pig in the small village herd was familiar to him—and, he related with rising passion, the smell of this particular pig was unmistakable! The offender already had a bad reputation and, without the curb of British justice, the errant swine would doubtless turn down a criminal path bringing strife to the whole village. Not a garden would be safe from the trespasser, he prophesied!

With a superior English air, the magistrate said that he was inclined to ignore Mr Serana's testimony as mere hearsay. Nobody had seen the pig, so nobody could know which pig had done the damage. 'Case dismissed,' he proclaimed in a loud voice, before ordering the courtroom to be cleared. Mr Serana had all the swagger of a man utterly convinced that his case was watertight, and as the verdict was translated he was rendered first incredulous, then furious, by this outrageous miscarriage of justice. He ranted and raved against the pig, its owner and the courts, who were clearly in cahoots, and vowed to have his satisfaction of them all.

With the court in uproar, the irate magistrate thundered, 'Another word, sir, and I'll hold you in contempt of court.' Stiff with fury, Mr Serana made his way between the school desks to the courtroom door. When under the lintel, inspiration came to him. He turned to face the magistrate, lifted his leg and let rip with a thunderous fart. This was all too much for the aged magistrate, who screamed out, 'That's contempt of court, sir. Three weeks in gaol for you.' At which Serana said politely, yet firmly, in pidgin English. 'What do you mean? You heard it, and you smelled it— BUT YOU DIDN'T SEE IT, DID YOU?'

With such tensions inherent in the justice system, it's inevitable that visiting magistrates occasionally fall foul of villagers who feel aggrieved at a perceived miscarriage of justice, and I was deeply dismayed to hear, some months after I had returned to Australia, that Mrs Shipley had been attacked by a man carrying a machete, and the family had been forced to leave the islands.

Working on Makira was challenging. The region around Sesena was formed of jagged limestone and was riddled with sinkholes. This made trekking into the bush slow, difficult and highly dangerous. One morning, while attempting to set a mist-net several kilometres from the village, I broke the rotting vegetation coating the rock and plunged into a sinkhole. It was a sickening feeling, falling in slow motion towards I knew not what. Perhaps I'd land on jagged limestone tens of metres below, or in a swollen underground river. Thankfully the sinkhole was only three metres deep, and I emerged from it shaken and covered in cuts, but otherwise unhurt.

Almost immediately, however, it became evident that the effort of setting the nets far from the village was worthwhile. At first light the following morning I trudged out, and found many bats hanging in the distant nets. Most were common species, but in one pocket hung a black bat with a body as long as my palm. Inspecting it closely, I saw that it bore bright pink spots on its ears and forearms. And when it opened its mouth I glimpsed four large canine teeth, each with a distinct groove running along its front edge.

It was a kind of blossom bat—a relative of the 'poison bat' that Sanila Televat had told me about on New Ireland. The blossom bats are an early offshoot of the flying fox family. They cannot echo-locate as do the insectivorous bats, instead they find their food—which is principally fruit and nectar—by sight. It would take me six years to collect enough blossom bat samples to identify and name the black blossom bat of Makira, and part of the funding for that work was provided by Emmanuel Fardoulis, an Australian who wanted to help environmental conservation. Ultimately I would name the species *Melanycteris fardoulisi* in his honour.

The study of the blossom bats revealed some fascinating insights into island biology. It turns out that every island in the southern Solomons has its own kind of blossom bat. Those inhabiting

Makira are not only the largest and blackest, but they also have the greatest difference in size between the sexes. So great is the difference indeed that at first I mistook the females, which lack the large, grooved canines and are much lighter in build, for a different species.

Several island species exhibit strange sexual differences, among them the huia bird of New Zealand, whose males have short stout bills while the females have long curved bills. Makira lies far enough south that its resources may be more limited than those of the other islands in its group. Rather than having the sexes compete for the same resources, species whose sexes specialise in exploiting different ecological niches may be favoured in such circumstances. But what do the males do with those enormous, grooved canines? My best guess is that they eat a lot of fruit, and use the canines not only to carry the fruit, but to puncture it and drain the juice. The females may be more strictly nectar feeders, visiting flowers for their food.

One morning some village boys shyly approached me holding a bagful of bats. They had collected them in a cave a little way from the village that they often harvested for bats to eat. The bag contained a kind of horseshoe-bat which had been long known from a single specimen collected in the nineteenth century and classified as a variant of one of the most common horseshoe-bats in the region. But the bats the boys had collected were strikingly different from any known horseshoe-bat in that the sexes differed in colour, the males being grey and the females bright orange. That, along with their small size relative to the more widespread diadem horseshoe-bat, was strong evidence that they should be classified as their own species. Following studies back in the lab I recognised them as another species unique to this increasingly intriguing island.

Meanwhile, areas nearer to home also were yielding surprises. Some boys from Sesena village had spied and captured a female flying fox that was roosting in a banana plant in a garden. She was carrying a single young and was covered in dense, tawny-golden fur. The only species she could possibly be, I realised, was the Solomons flying fox, which had been recorded from Uki Ni Masi in the nineteenth century. But, as with the horseshoe-bats, she was so different from flying foxes found elsewhere that she too had been misclassified. Now known as the Makira flying fox (*Pteropus cognatus*), the species is part of a growing list of fauna recognised as being unique to Makira and its nearby islands.

For all of the success I was having with bats, I'd not caught sight of the object of my search—the mysterious rat of Ugi (or Uki Ni Masi). I questioned villagers closely about rats, and it was clear that they knew of both large and small kinds. I was somewhat handicapped, however, by not knowing what the creature looked like. At first, villagers said that large rats were everywhere, but it turned out that they were referring to the common black rat, which had arrived on the island when a trawler sank there in 1984. This rat is indeed larger than the common (and also introduced) Pacific rat, but it was not what I was looking for. Then I discovered that a few people knew of an even larger kind of rat. It was restricted to the rugged inland plateau and ranges of the island, which rise to around 1000 metres above sea level. This creature may have been my mysterious rat but, by the time I'd sorted all this out, the expedition was coming to an end and I needed to return to Australia.

For twenty years I thought that I'd blown my chance of seeing the mystery of the identity of the giant rat of Ugi solved. But then, in 2009, an ornithologist sent me a photograph of some rat footprints in a muddy puddle. He had taken the picture whilst birding high in Makira's mountains, and the tracks had without doubt been made by a giant rat. They were, apart possibly from the mystery skull collected in 1881, the first solid evidence that the giant rat of Makira actually exists. As I write I have a student who is on the track of this elusive creature. Perhaps I'll yet live long enough to see the one-hundred-and-thirty-year-old mystery solved.

The small aircraft that carried me back from Makira set out for Henderson airfield on Guadalcanal at dusk. Flying away from the red rim of the sunset into the engulfing dark of the Pacific night, I could imagine the feelings of a bat or bird blown off course over the ocean or a rat stranded on a floating tree. For them, nothing but a lucky current carrying them towards a new land and their own ability to endure could save them. Despite the fact that we had navigation equipment aboard our small plane, I felt scared once the last few lights of Makira slipped past us, for ahead there was nothing but an ocean of blackness. What would happen if we somehow missed Henderson Field? Would we too be swallowed up by that blackness, like countless birds, bats and other creatures that were blown or wandered off course in aeons past? Thankfully we did arrive safely, but the experience left me with a new respect for the vastness of the Pacific and for those species that have conquered it.

CHAPTER 11

Malaita

I'd heard from Mike McCoy that he planned to visit one of the most inaccessible places in the Solomons—the land of the Kwaio on the island of Malaita. When he suggested that we make the visit together, I jumped at the opportunity. The chance had arisen when Mike met Simon, the son of Folofo'u, an important Kwaio man now living in Honiara. Mike had given Simon his Swiss army knife, and the Malaitan had reciprocated by inviting him to visit the hamlet of Naufe'e in eastern Malaita. Malaita is the most densely populated island in the Solomons archipelago. Even its mountainous spine, which rises to around 1000 metres, supports a fair density of settlements. But Naufe'e lies in the most inaccessible part of the mountain chain. There was a chance that primary forest, and therefore unique fauna, still survived there.

The great tub of an inter-island ferry that makes the crossing from Honiara to Malaita's administrative centre of Auki normally takes only half a day. But on the trip we made the vessel broke down repeatedly, drifting helplessly for hours at a time in an increasingly foam-whipped sea. On each occasion it was a relief to hear the great diesel engines throb back to life, as the risk of being blown onto a hidden reef or rocky shore was very real, and it was far from certain that our overcrowded ferry had sufficient life rafts and vests for all. At last the tiny harbour of Auki hove into view. From there we would cross the island by road, then travel by canoe into Sinalagu Harbour—the gateway to the Kwaio territory—on the east coast.

The Kwaio had remained cut off from the world. They were a proud and independent people, and the mountainous centre of their homeland was almost entirely inaccessible to the colonial government. From that jumble of abrupt mountain ridges, valleys and dense jungle, warriors would sally forth on raids against their neighbours. Their most powerful and influential leaders were known as *ramo*. Each a combination of strong-man and bounty hunter, their prestige and wealth depended upon collecting and distributing blood money. Such bounties were frequently offered by the relatives of women who had been seduced in violation of the island's strict sexual code, and the *ramo* would sometimes even kill their own kin and followers to get a payment.

By the colonial period *ramo* were using antiquated Snider rifles, often firing them at point blank range. Examples held in museum collections reveal the esteem in which they were held. Decorated with intricate shell inlay, these functional firearms had been turned into extraordinary artefacts. To the *ramo*, treachery was a way of life. Inviting a victim to a feast and, when he was relaxed, pressing the barrel of the Snider to his ribs and pulling the trigger was a favourite technique.

To be recognised as a *ramo* you had to kill a person with a bounty on his head in personal combat. Then, with a great show of bravado, the *ramo* and his henchmen would storm into the settlement that had offered the bounty and collect the payment of shell valuables from a pole in a clearing for all to see. Frequently, relatives of the slain man would offer another bounty, and so the vicious cycle would continue. The Kwaio *ramo* were the toughest of the tough, and by the early twentieth century they were killing blackbirders (labour recruiters for Queensland sugar plantations) and missionaries alike. As an anthropologist explained, at the time it was 'tremendously dangerous for any European to land on Malaita or expose himself to attack in any way.'[14]

So it was that Malaita came to be the ragged edge of what was just about the most remote and forgotten part of the British Empire. Yet such attacks could not be tolerated for long. Sporadic, failed attempts were made to extend the *Pax Britannica* into Malaita's mountains and by the mid-1920s things were coming to a head. The British colonial administration began collecting taxes in the area, and the *ramo* understood that this was a direct challenge to their authority. One *ramo*, a man named Basiana—arguably the most fearless of them all—who lived in the mountainous Gounaile area, was determined to do something about it.

It fell to district officer William Bell to undertake the 1927 tax collection. On Tuesday 4 October he, his European assistant Lillies, a Malaitan clerk called Masaki and thirteen armed Solomon Islands policemen occupied a crude tax-collection hut in Sinalagu Harbour and waited for the mountaineers to deliver their payments. Bell, an Australian, was a highly experienced administrator who knew the Kwaio and their ways well. He understood that this would be a test of strength, and he had deliberately come ashore, rather than collecting the taxes from his boat, for to do otherwise would be a sign of weakness. When the hundreds of fully armed Kwaio males arrived in the clearing before the tax hut Masaki said, 'They've come to kill us. I can see the blood in their eyes.'[15] Bell's response to the crowd was coolly professional:

> I've come to collect taxes today. My police say they can see
> you've come to fight with us today. But I don't want to fight
> with you. I told them if you wanted to make trouble with
> us you'd have to start it yourselves. We've come in peace.[16]

While one of Basiana's men distracted the Malaitan policemen with a string of very valuable old shell money, others were stealthily cutting through the cane fastenings that held the wall panels of the tax hut in place. Then Basiana stepped forward, paid his tax and quietly walked away. As his henchmen lined up to pay, Basiana strolled back to the edge of the clearing where he'd laid down his pouch, and he stealthily removed from it the barrel of his Snider rifle. It had been consecrated to his ancestor Ma'una, and was full of spiritual power. Basiana concealed the barrel between his arm and body and walked once more towards Bell, who remained at his desk receiving payments. Basiana got to the head of the queue as Bell was looking down, writing in a tax roll. Swiftly raising the barrel he struck Bell's head, which exploded with an awful sound, spattering brains and blood everywhere.

Almost instantaneously another warrior struck Bell's assistant Lillies. But it was only a glancing blow to the skull, and Lillies rallied. Then Basiana jumped onto the table and rushed into the tax hut, knocking aside the police rifles before they could be used. With its stays cut, the hut was collapsing, trapping Lillies and the police who were inside. Within seconds, two Europeans and thirteen Solomon Islanders lay dead. It was an overwhelming victory by a force armed largely with traditional weapons against colonialists equipped with the latest rifles. Only one attacker had been killed and half a dozen wounded. Yet the massacre would bring cataclysm to the Kwaio world.

The dead would later be buried by a revenge party beside Sinalagu Harbour. But Lillies' left hand, which was severed by Fenaka of Ailai, was never recovered. It was, the Kwaio say, taken into the bush, where it was smoked and kept as a talisman. As far as anyone knows, it remains in Kwaio possession to this day.

By 10 October, just six days after the massacre, the cruiser HMAS *Adelaide*, carrying nine six-inch guns and numerous smaller armaments, was on her way to Sinalagu Harbour. The Kwaio must have watched in awe as her huge bulk steamed into view. Seeing photographs of the great vessel at her destination, it's impossible to avoid an overwhelming sense of European power. The vessel landed tonnes of supplies and a large punitive force, but the Europeans were almost entirely ineffective in the steep terrain and dense bush. So another approach was tried: arming the Kwaio's enemies.

These people lived adjacent to the Kwaio and had suffered under the Kwaio *ramo* for decades. They knew the land, their enemies and their ways, and were thirsting for revenge. They also knew that the best way to destroy the Kwaio was to desecrate the ancestral shrines. They smashed ancestral skulls, or tossed them into the women's menstrual huts. The bark mats upon which menstruating women sat were draped on the sacred sacrificial stones. Ancestral relics and other sacred objects were burned, the intention being to bring down the wrath of the ancestors on the Kwaio. Although it was forbidden by the Europeans, Kwaio men, women and children were massacred wherever they were encountered. It was said that for every two prisoners the revenge parties brought down from the mountains, one Kwaio was shot. To this day the full tally of dead is unknown, but it seems that at least sixty Kwaio died in the weeks after the original massacre.

Basiana surrendered and was taken to the colonial head-quarters at Tulagi where, on 29 June 1928, he was hanged in front of his two sons, fourteen-year-old Anifelo and his younger brother Laefi. Before climbing the scaffold Basiana placed an ancestral curse on Tulagi, saying to the resident commissioner and comman-dant of police, 'Tulagi, where you have your flag, will be torn apart and scattered.'[17] Taken as something of a joke at the time by the Englishmen in their club, it was a curse perhaps remembered by Solomon Islanders when the Japanese drove the British from Tulagi fourteen years later.

The impact of these events on the Kwaio were long-lasting. As Riufaa of Kwangafi said:

> When they destroyed our shrines, they destroyed all the good things in our lives…Everything was bad after that… The ancestors' consecrated pigs had just been eaten, the sacred things had all been defiled—how could our living be straightened after that? How can we live at all? We are finished.[18]

But the Kwaio were not finished—just hardened and deter-mined to turn their backs on the world. They repulsed all efforts at Christianisation—their mountain fortress was an inviolate retreat where the old ways would persist. As late as 1962, when anthro-pologist Roger Keesing and his wife first went to live among them, he was warned that his life would be in danger—a warning only amplified when a missionary in the area was speared to death soon after the Keesings arrived.

It was into the heart of this world that Folofo'u's son had invited us. He was an impressive man—all two metres of him—with a broad chest and bearing that would do any sergeant major proud. When he met us in Sinalagu Harbour on that bright morning I got the sense that here was a person the outside world had not yet bowed. As we set off up Sifola—the coastal slopes lying behind Sinalagu Harbour—the heat was stifling, and Mike and I were soon puffing and sweating under the weight of our packs. Having walked for several hours we reached a clearing at an elevation of around six hundred metres, and we rested. Then an amazing thing happened. A young woman, puffing on a homemade corn-cob pipe and dressed in nothing at all, came striding down the path towards us. She casually took my overstuffed pack from my shoulders and set off up the slope. For a moment I wondered whether heat stroke was taking its toll or whether I had died on that kunai slope and gone to Kwaio heaven.

I was roused from my reverie by three similarly un-clad young women, none of whom were lucky enough to possess a corn-cob pipe. As they passed they flashed brilliant smiles, and went on to relieve Mike and our Kwaio helpers of their packs. I followed the young woman with the corn-cob pipe. She puffed clouds of smoke as she strode up the hill with my heavy pack on her back, her pace putting me to shame. As I struggled upwards I turned to Mike and asked why the young women were naked. 'Kastom bilong Kwaio'— just the way they do things up here—he replied. The unmarried Kwaio females had never worn any clothes, he added, but when they married they donned a tiny patch of material to cover the pubic area—that was all.

The Kwaio set me thinking about how it could be that the

women could be so relaxed about being naked. I never got to know Kwaio culture sufficiently well to answer that question, but it might have something to do with Kwaio sexual morality, which is exceptionally strict. Even at the time of my visit, a man who seduced any female was likely to wake up with a spear through his ribs. Perhaps the protection this affords young women contributes to their unconscious attitude towards nudity, as, doubtless, would the idea that nudity is the socially sanctioned condition for them.

Soon we came to an abrupt plateau lying six hundred near-vertical metres above the glorious harbour of Sinalagu. The crisp atmosphere seemed to magnify my sight so that even the tiniest details of the hamlet we had left hours ago could be made out below. Naufe'e, which sits on the plateau, remains unique in my experience of Melanesian villages. Everything was built of bush materials, the only concession to modernity being a pipe and shower-rose set up in a clearing where the children would wash each day.

At one end of the hamlet stood a men's house, with two tree-fern poles at its rear, each one with a niche carved into it to hold a human skull. Regrettably I was never allowed into the house or allowed to photograph it. The best bush was also taboo, for those glades were dedicated to ancestral shrines and had to be left alone. In the middle in an open area were the dwellings, and below the houses were the women's menstruation huts and latrines. It's a layout that was typical of villages in many parts of Melanesia in pre-Christian times, yet I had rarely seen it, for even in areas that missionaries had failed to reach, Christian ideas had permeated, resulting in the destruction of traditional culture.

Naufe'e had its chief in the form of Folofo'u, who was in his eighties or nineties when I met him. He had been a young man when Bell was killed and had somehow survived the massacres. I am sure it was his influence that had kept Naufe'e isolated and one of the most traditional places in the Pacific. But change was coming. The very fact that Folofo'u's son had invited us testified to that.

That evening I went hunting with Simon. He towered over me, shotgun in hand, as we scrambled along muddy paths, seeking flying foxes and possums. It was a great night to hunt. The sky was clear and it had not rained for several days, which meant that the blossoms of the Malay apple trees were full of nectar. We bagged several possums and three flying foxes, then returned to Naufe'e in the small hours to sleep.

By the time I awoke the sun was high in the sky. A young woman was sluicing her body under the communal shower, the sun catching her curves so that she looked like a classical nymph. After a quick breakfast of hard biscuits and tinned fish I set to work weighing, skinning, measuring, sampling, recording and preserving the flying foxes and possums Simon and I had bagged the night before. It was important I do this quickly, for Simon wanted their meat for food, and in the tropical climate it would soon spoil.

Skinning and sampling is hard work, and as usual in a Melanesian village I attracted a crowd of onlookers—in this case young women and girls. Wanting to see precisely what I was doing, they pushed forward in a crush, and before long I felt a firm breast resting on my shoulder and the warmth of a groin pressing into my side. This made it difficult to concentrate, and after I'd nearly pushed the scalpel blade through the palm of my hand I begged Mike to distract the crowd. He got out his instamatic camera and began clicking off prints. As he handed them about my audience was delighted, and naturally gravitated to him.

I was determined to hunt again that night, but in the evening the rain set in. Despite a long slog through the sodden bush we saw no more flying foxes, for the rain had washed the nectar from the blossoms and the bats had gone elsewhere for food. I also discovered that, despite its isolation, the forest surrounding Naufe'e was mostly disturbed. There were few if any trees venerable enough to serve as a roosting site for a monkey-faced bat. Indeed the local people had no knowledge at all of such creatures. If these bats had ever been present on Malaita, they are now presumably long extinct.

The best way to find out about the region's fauna, I realised, was to sit down with Folofo'u. He had an encyclopedic memory, and listening to him was like travelling back in time. In his youth Malaita had more primary forest than it does today, but most of it was cut down while he was young. He recalled that when he was a child, his father had caught a large, forest-dwelling rat with a tail like a rasp. It could only have been a member of the curious rodent genus *Solomys*, a group of giant rats unique to the Solomon Islands. There is a species on every major island except Guadalcanal (which has its own *Uromys* rats) and Malaita. The absence of giant rats from Malaita had long been a mystery. But Folofo'u had solved it for me. There *had* been a giant rat of Malaita, but it had become extinct, perhaps through deforestation or the introduction of cats, within living memory.

CHAPTER 12

The Face Never Seen

By 1990 I felt that I had enough experience to attempt the high peaks of Guadalcanal. There had been no softening of attitudes among the landowners at Gold Ridge, which left me just one option: ascending Mount Makarakomburu via the weather coast. The trouble was that the political situation on the weather coast was deteriorating as local warlords gained power. The most fearsome was Harold Keke, who would go on to become the self-styled head of the Guadalcanal Liberation Front. In 2002 he murdered Father Augustin Geve, a Catholic priest and a member of the Solomon Islands parliament. The following year one of Keke's henchmen, Ronnie Kava, would be accused of murdering seven Anglican clergymen.

Although these events were still some years in the future, it was clear that life was already cheap on the weather coast, and that the longer I delayed the expedition the more dangerous it was likely to become. By May 1990 I could put it off no longer. Mike

McCoy was occupied with other work, so I teamed up with Tanya Leary, who was then working for the Solomon Islands Ministry of Conservation. She has gone on to make a fine career for herself in wildlife conservation in Melanesia.

Tanya and I reckoned that the best way to reach the weather coast was to follow the road west of Honiara as far as it went, then to try to travel by sea to reach the area. And so it was that we found ourselves at the end of the road, quite literally, in Lambi village near Guadalcanal's eastern tip. The weather was foul—blowing a storm and raining incessantly. But, not to be put off, we hired a tiny canoe to take us from village to village in search of a sturdier vessel with a captain willing to make the journey. As we travelled through rough seas and pelting rain the canoe almost filled with water threatening to drown us. By dusk it was clear that, for this season at least, nobody would risk his life by making the trip.

Frustrated and sodden, we saw no alternative but to return to Honiara. There was just one last possibility. A helicopter used by the gold mining company at Gold Ridge was temporarily based at Henderson Field. If we could charter it, we might be able to get dropped off in Valearanisi village, from whence we could climb the mountain. At an estimated cost of $1100 it was an expensive option, but it was the only one remaining, so after making the arrangements Tanya and I boarded the chopper. On our way in we flew over the summit of Mount Makarakomburu (which was entirely wreathed in cloud) and on to the rugged and wet weather coast of Guadalcanal. As we descended between steep, vegetation-covered ridges and banks of cloud, I could see our destination ahead—a small collection of huts perched on the flank of the mountain, beside which ran one of the most amazing scenes of destruction I have ever laid eyes on.

Valearanisi is located at a few hundred metres elevation on the eastern side of Makarakomburu. Beside it ran a vegetation-free scar—at least a kilometre wide and the length of the mountain-side—which was filled with boulders, many larger than a house. It was once the gentle, soil-covered valley of the Kohove River where village gardens flourished, but Cyclone Namu, which had struck the region four years earlier in May 1986, had gutted it, leaving a hundred people dead and thousands homeless. Judging from the massive erosion scar, its impact on the forest ecology of the weather coast of Guadalcanal must have been catastrophic.

The people of Valearanisi were expecting us, as we had managed to get a message through that we were coming. The locals, as Peter had indicated, were friendly and hospitable. But they had learned well from the mining companies, and were expecting a $500 fee just to permit us to step onto their land. The cost of the helicopter, we discovered, had doubled to in excess of $2000. With our limited budget, the villagers' demand for a cash payment threatened to derail the expedition.

That, however, was only the beginning. I was soon served with a comprehensive employment agreement that would have bankrupted us. Anyone who worked with us, it stated, must receive, in addition to daily pay, special night-time loadings, time and a half on Saturday, Sunday off and a living-away-from-home allowance if they left the village. Where a century before Woodford had battled headhunters, I found trade unionists of such sophistication that they could teach the Australian Council of Trade Unions a thing or two. Underpinning it all was the threat of Keke's thugs. Having gone so far, however, there was no question of turning back. We had to pay the fees, with anything in excess of the grant coming out of my own pocket.

The morning following our arrival broke brilliant and clear, with the summit of Mount Makarakomburu seeming so close I felt I could almost touch it. But such proximity is illusory. In the formidable country of the weather coast, travelling just a few kilometres can take all day. Burdened with hundreds of kilograms of equipment, including liquid nitrogen cylinders, nets and traps, it took us two days of exhausting toil to reach just 1200 metres elevation. With the weather deteriorating and the mountain summit looming more than 1000 metres above us, a decision had to be made. In such circumstances one either walks or works—it is impossible to do both as there is simply not enough time to set up and collect nets and traps as well as make camp. We had passed a flattish area at around 900 metres and with the terrain swiftly becoming more rugged with greater elevation, we decided to make it our base camp. From it we would make day trips to higher regions—and perhaps even reach the summit.

Just a few hundred metres beyond the village the vegetation was magnificent, undisturbed mountain forest, the character of which changed continually as we climbed. At around 1200 metres an elegant palm, festooned with red fruit, became a dominant plant. This graceful tree stood out above the canopy, forming a distinct band on the mountain and marking an ecological change that looked promising in terms of fauna.

We quickly set up mist-nets around the 900-metre camp, and almost immediately started to catch interesting creatures. One afternoon there was a large, blackish honeyeater with striking yellow side-plumes. It was one of the Solomon Islands' most special birds—a honeyeater that bears the marvellous scientific name of *Guadalcanaria inexpectata*. A genus and species that is entirely restricted to the high mountains of Guadalcanal, it must be descended from ancestors that made their way to the islands millions of years ago. Known from just a handful of museum specimens collected in the 1920s, it had not been seen by biologists for many years. Our record confirmed its survival and so was an important contribution to conservation.

The next day I set out early with two boys from the village, hoping to climb high up the mountain to set mist-nets. We planned to wait and check them that night. As we climbed to around 1700 metres elevation, the vegetation underwent an abrupt change. The dense mossy tangle gave way to more open undergrowth, and a trunkless palm became dominant in the understorey. The tall, red-fruited palm had dropped out way below, so here was evidence of another marked ecological change—and with it the chance of a change in fauna. The trekking was so tough, and the weather so appalling, however, that we had to abandon our plans of staying out all night. It was with great regret that I made the decision to turn back and work at lower elevation.

On the way up, at around 1230 metres, we had found a saddle between two peaks—an ideal place to set a mist-net as such places are often used by bats and birds as a fly-way from one valley to another. We stopped there on the way down, set up a mist-net and laid our rat-traps, then rested by the net, waiting for dusk. A little while after the sun fell below the horizon I heard a large bat land on a branch by the net. It was about ten metres away so I couldn't see much detail, but it seemed different from any species I'd seen before, with long fur and a rounded head. It only stopped for a moment before flying off, evading the net and leaving me intrigued.

That evening as we hunted on our way down the mountain to base camp, I was disappointed that we had seen so little. Then, at around 10 pm, I saw a bright pair of eyes in the forest ahead. I raised the shotgun and fired, and I was amazed to find that they belonged to a huge, mean-looking black cat. This was a dismaying discovery, for cats are one of the greatest destroyers of island life. If feral cats had penetrated the mountain forests of Guadalcanal in any numbers, there was little hope for the survival of emperor, king or little pig. The local people later confirmed that the great rats were already very rare or extinct in the region. The old man called *Hue Hue* that Peter had told us about was, needless to say, conspicuous by his absence.

The following morning villagers arrived carrying the mist-net and traps we had set on the mountain the day before. At first I was delighted, as the traps all seemed to hold rats—big black ones that I did not immediately recognise. Yet another blow to our efforts was the discovery that they were merely huge, melanistic (blackish) examples of the introduced Pacific rat. The whole terrestrial fauna of the place, it seemed, had been taken over by the duo of intro-duced cat and introduced rat, both of which had become large and black-furred, perhaps as a result of the dense vegetation and cold climate. This was as depressing as it was unexpected.

After swallowing this bitter pill I began opening the canvas bags that held the bats caught in the mist-net. There were a surpris-ing number of them, indicating that the saddle saw a lot of bat traffic. The smaller bags held familiar species, but as I opened the largest bag, which I left to last, I saw a new face peeking out at me. It was the eyes that first struck me. They were a deep ochre-red—a shade that I had never seen in an animal's eyes before.

Gingerly lifting the bat from its bag, I found that it was a creature around half the size of a common Australian fruit bat. It was in many ways utterly unique. It had a cutely rounded head with a stubby snout and short ears hidden in long black fur. The body, in contrast, was covered in luxuriant fur of a golden-khaki colour. It was the wings, however, that stood out: while ordinary-looking and black on top, their underside was strikingly marbled in black and white, and unlike any bat's wings ever recorded. I showed it to the villagers who were with us. They included some experienced hunters, but none of them had ever seen a creature like it and they had no name for it. Upon reflection this was hardly surprising, as the people of Valearanisi rarely visited the upper slopes of Mount Makarakomburu, which to them is the haunt of spirits, and they did not hunt there at night. Having examined the great mammal collections of Europe and North America, I knew as well that nobody had ever collected anything like it previously.

New species of mammals are still being described today, but most are small and obscure—mice and small insectivorous bats tend to predominate. And when a larger species is described, it's often based upon a museum specimen that has somehow been overlooked by earlier researchers. It's rare for a new species of largish mammal to be discovered in the wild. Even then the discoverer will usually find that the local people have long known of its existence, and that the only glory he can claim is to announce its existence to the world via a scientific publication. But as I held that bat it slowly dawned on me that I'd stumbled into one of the rarest of circumstances. That morning I'd had the privilege of peering into the face of a mammal hitherto unknown to everyone. In all my years of fieldwork it was the only occasion I would make such a completely novel discovery.

While the precise identity of the bat puzzled me, it was immediately clear as to which group it belonged. It was a miniature and highly ornamental monkey-face. While the lowlands were home to its larger and funereally black relative, conditions high on the mountain had selected for different traits, giving rise to this pygmy form. When I returned to Australia I set about writing a description of the new creature and giving it a name. I settled upon *Pteralopex pulchra*, meaning the beautiful winged Arctic fox. To this day nothing more is known of the species than the few notes I made on that morning while high on Makarakomburu over twenty years ago.

As the days wore on the villagers seemed to get greedier and more offensive, asking obscene prices for food and assistance. As a result, our relationship with them soured and I took to sleeping next to my shotgun and machete. Not that they'd be much use to me if things unravelled completely. And on top of it all I had to face the fact that I'd never see the summit of Mount Makarakomburu. Time was too short and the situation too volatile to contemplate an ascent. As the confrontations with the locals worsened I found myself frequently on the brink of losing my temper. This is a rare thing for me, and I realised that we needed to leave the mountain as soon as possible if we were to avoid a great unpleasantness.

Over the years I have frequently thought about the behaviour of the people of Valearanisi village. One factor that affected them was that the place had never been brought under full government control, so the old ways persisted. And there's no doubt that the goldmine and the tales of enormous wealth being extracted from it affected the villagers' view of Europeans. Although they lived on the far side of the mountain, they expected a share of the wealth, yet they received nothing. Not surprisingly, many saw the mining as theft of their inheritance.

Perhaps the situation had been compounded by the generous assistance provided to the people of Valearanisi by the Australian government in the wake of cyclone Namu. At first I thought that this might spark reciprocity, but they saw things very differently. The people of Valearanisi were, I'm now certain, expecting me to emulate my countrymen and dispense more of my apparently limitless wealth. When I disappointed them, anger and frustration boiled to the surface.

The entire social situation was made worse at the time of our visit by a bizarre millennial cult. Some young men had travelled to Honiara where they had seen a propaganda film, made by American Christian fundamentalists, depicting the Rapture—the end of the world. As portrayed in the movie, the Rapture will see the righteous miraculously levitate to heaven, leaving the damned to perish on a godforsaken planet. The young men who had seen the movie graphically described to me the coming fate: aircraft falling from the sky, terrible car accidents and other catastrophes. With great conviction the village head told me that the world would end in 2000—then just a decade away. I discussed their beliefs for hours, but all my efforts to persuade them that the Rapture was fantasy were dismissed. Moreover, I thought their greed strangely dissonant with this belief in the end of days.

We had appointed a day for the helicopter pilot to pick us up, but I was increasingly concerned that deteriorating weather would prevent its return. Nothing is certain in Melanesia, except the transience of life, and by then relations had frayed so badly that the evening before the chopper's anticipated arrival was one of the most tense I've ever experienced. Winifred Violet Scott's generous bequest was all but exhausted, yet I was still being bullied and cajoled, almost incessantly, for exorbitant payment for necessities like shelter for the night.

How to explain the frustration that saw me pick up my bush-knife in anger? Yes, I was worried that someone might steal our equipment, but it was blood—stupid youthful blood—that had me thinking of acting in ways that shame me even now. I thank the god of good fortune that I could contain my anger long enough to hear the dull thudding of the helicopter. Within the hour we were on our way to a completely different world. For months after, Tanya told me, villagers from Valearanisi would visit her office in the ministry of conservation seeking compensation payments for our visit.

As the sound of that thudding skyhook reverberated throughout the cyclone-devastated valley just before we left, I received a wondrous and entirely unanticipated gift. An old man, whom I had met when I had first arrived, hurried towards me, a plastic bag in his hand. I had explained my interest in bats to him, and he offered me the bag with a look of puzzled expectation. In it was a tiny winged creature with silver-tipped, blackish fur as beautiful as the fur of any chinchilla. It was an almost mythical animal—the flower-faced bat (*Anthops ornatus*) of the Solomon Islands.

The species had first been collected by Charles Woodford, in 1886, but the last time it had been seen was when my predecessor at the Australian Museum Ellis le Geyt Troughton received a specimen from Father Poncelet, a Belgian Catholic missionary based on Bougainville, in 1936. I peered into the bat's strange face. Its eyes were completely hidden in its luxuriant silver pelt and a strange, bright orange, flower-like growth covered much of its face. Predominant in this growth were three tiny, perfectly spherical orange balls, lined up across the forehead and, below them, numerous overlapping flaps of almost luminous orange skin that formed a sort of radar dish around the nostrils.

What trick of nature, I wondered, had formed this living jewel perhaps ten million years in the making? The flower-faced bat's ancestors had most likely flown to the Solomon Islands some time before North and South America were joined by the rising isthmus of Panama, before the island of New Guinea had formed into its present shape and long before anything like an upright ape trod the earth. Indeed, I later discovered, its nearest relatives are fifteen-million-year-old fossils from Australia.

What had inspired this very bat—which as far as we know is forest dwelling—to fly into the house of an old villager who just happened to know that there was a white man in the area who would appreciate this gift? He told me how he closed the shutters of his house and he and his children chased the creature about before catching it, and I handed over the very last of Winifred's cash and accepted this most marvellous offering.

Honiara's Mendana Hotel—with its cold beer and hot and cold running water—seemed a strange place to be. Just hours before I had stared into the depths of my own potential evil and then received what seemed like a gift from the gods on that worlds-away weather coast of Guadalcanal.

CHAPTER 13

Poncelet's Giant and the
Last Great Forests

At its northern end, the archipelago that is the Solomon Islands terminates in the mighty island of Bougainville and its outlier Buka. With mountains nearly 2600 metres high (just taller than those on Guadalcanal) and with an area of eight and a half thousand square kilometres, Bougainville is a biological wonderland that is still yielding unknown species. But I have never worked there. The island was well serviced with airports and roads, and it just looked too easy. So I thought Bougainville could wait. How wrong I was!

The Panguna copper mine on Bougainville, which began operations in 1969, had presented the local people with a host of problems, including pollution, social disruption and disadvantage, none of which were being adequately addressed. In 1988 a Bougainvillean called Sam Kaona, who was frustrated at the lack of progress in addressing the problems, took to the jungle and formed a ragtag militia that was to become known as the Bougainville

Revolutionary Army. It was not taken seriously at first, but by May 1989 Kaona had forced the Panguna mine, one of the largest and most profitable in the region, to shut down. Then followed a long and brutal civil war, now thankfully brought to a close. But to this day the mountains of southern Bougainville remain too dangerous to visit.

Although Bougainville was off-limits I was able to learn something of its fauna through an archaeological dig that had taken place on the adjacent island of Buka. During times of low sea level 20,000 years ago, Buka was joined to Bougainville, so they share a common fauna. An archaeologist at the Australian National University, Matthew Spriggs, had excavated a cave that preserved the remains of animals dating back to the last ice age. I had time to do a detailed study of only the rodents, but it revealed five species of large-to-gigantic rats, two of which had never been seen alive. Both appeared to be ground dwellers, like the emperor rat of Guadalcanal, and one was the largest rat ever recorded in the Solomons archipelago. Their remains occurred throughout the deposit, so perhaps relict populations survived. But how was I ever to follow this lead with a civil war raging on Bougainville?

Buka and Bougainville were not the only islands joined during the last ice age. Twenty thousand years ago low sea levels meant that most of the northwestern islands of the Solomons chain were united in one vast landmass known as Greater Bukida. It must have been one of the largest islands in the Pacific, encompassing every land from Buka to the Florida Islands just off Guadalcanal. Back then, animals roamed freely throughout this great landmass, meaning that, in the lowlands at least, the fauna of all the islands should be similar. Perhaps the scattered fragments of Greater Bukida offered a means of investigating Bougainville's fauna.

One species recorded from the fossil deposit on Buka had been observed just once by a European. Known as Poncelet's giant rat, it had been collected by the same Father Poncelet who had sent Ellis Troughton the flower-faced bat. Poncelet was based at Buin, in the south of Bougainville in the 1930s. During a visit to Sydney he had called in at the Australian Museum to enquire whether he might assist the scientists by collecting samples. His offer was gratefully accepted, for even at that time travel to the mountains of Bougainville was dangerous and rare. As noted in a 1935 newspaper report, a missionary (perhaps Poncelet himself) had set off in search of an unknown community which was rumoured to live in the mountains. He was armed with only a walking stick, and when he located the settlement the natives were so amazed that they:

> actually rubbed his skin to make certain the colour was not painted on. Discussing his fate, the natives apparently concluded that a man of peace was not worth killing, to the relief of the priest, who said he was naturally terrified, as the natives, though not cannibals in the sense of killing to eat, have been known to kill and feast upon intruders.[19]

To the delight of Ellis Troughton, Poncelet dispatched a carefully documented collection from Bougainville soon after his visit. Its outstanding novelty was a pickled body and two skulls of a prodigious rodent hitherto unknown to science. The creature was the size of a cat, had a naked, prehensile tail and was covered in very coarse dark chestnut hairs that formed a kind of mohawk along the head and neck. Father Poncelet's accompanying letter said that they were found ten miles from the mission station in dense forest, were known to the natives as *nagara* and were very rare. To honour the redoubtable Belgian, Troughton named the creature *Unicomys ponceleti*.

Several zoological expeditions have visited Bougainville since Poncelet's day, yet nothing more was heard of this extraordinary rodent. Was it, like the emperor rat of Guadalcanal, already extinct? Choiseul and Isabel are, with the exception of Bougainville, the largest fragments of Greater Bukida, and being politically part of the nation of the Solomon Islands they were not drawn into the civil war. They represented our best chance of seeing Poncelet's giant rat and the rats that might still occur on Bougainville. The survey team included Harry Parnaby, an Australian expert on bats, Ian Aujare, a Solomon Islander with a keen interest in wildlife, and Tanya Leary. As Ian was from Isabel, he was able to open many doors for us with local communities.

As a result of disease, blackbirding and headhunting raids by the inhabitants of New Georgia Island, Choiseul and, more particularly, Isabel had been almost depopulated by the early twentieth century. Even at the time of our work their populations remained small, and both islands sheltered tracts of magnificent lowland jungle. We suspected that it was in such jungle, where the rainforest trees reach their greatest girth and height and where a great variety of fruits and nuts abound, that Poncelet's rat and its kin might have found a haven.

After weeks of fruitless searching the team finally met a hunter who claimed that he knew of a great rat. It was not called *nagara* as on Bougainville, but *vusala*. But was *vusala* Poncelet's giant rat? The only way to be certain was to examine one. The hunter searched in a large swamp forest near the village of Vudutaru, and there, in February 1990, he located a female accompanied by a single young almost half her size. They had been sleeping together in a huge messy nest made of sticks resembling an eagle's nest, perched high in one of the largest trees in the forest.

When the specimens reached the museum I compared them with those collected by Father Poncelet sixty years earlier. One difference was immediately evident. The hairs of the Choiseul animals were black, rather than the rich chestnut of Father Poncelet's rats. But did that mean they were a different species? Comparisons of teeth, bones, feet and other anatomy indicated that no other significant characteristics existed. The difference in colour, I concluded, must have resulted from the preservative used by Poncelet. We never could identify precisely what chemical had altered the colour of the hair, but at least we could say that, after an absence of nearly sixty years, the giant rat of the north Solomons had been rediscovered.

The information the expedition collected on Choiseul specimens is vital to the conservation of the species. First and foremost it demonstrated that this almost legendary rodent still survived. Secondly it indicated that its reproduction was slow: just a single young was reared at a time, and the young stayed with the mother until it was nearly mature. Even moderate hunting pressure could drive such a creature into extinction. But our work also revealed that the species requires pristine, unlogged forest—unless the unethical loggers who are devastating the region are stopped, the creature cannot survive. Tragically, Poncelet's giant rat is not the only biological treasure under threat. As we were soon to learn, the monkey-faced bats, smaller giant rats and many lesser creatures of the Solomon Islands require unlogged forest for their survival.

The distribution of animals on the Solomon Islands makes no sense unless you take account of the ice age. Some islands that are now widely separated were then joined up, while other islands lying adjacent to each other remained separate because deep channels persisted even when the sea fell to its lowest point. One region has always remained isolated. Its largest islands are Kolombangara, New Georgia and Vangunu, and the waters, reefs and coral cays of the region are pristine, making it one of the most beautiful places in the southwest Pacific.

The only native land mammals reported from the region were bats, so I assigned Harry Parnaby, the bat expert on our team, to head up the survey of the area. One specimen he collected would launch a new research project. It was a small monkey-faced bat— the smallest yet known—with bright orange eyes, mottled wings and short golden fur. Clearly an unknown species, it was nonetheless evidently related to the monkey-face I'd found on Mount Makarakomburu. Harry would go on to name it *Pteralopex taki*, *taki* being the name that the villagers of New Georgia and Vangunu knew it by. It's a great tradition, I think, to use the local language name for species new to science. It acknowledges the expertise of the local people in regard to their fauna, and such names do gain global acceptance. After all, that's how we got the likes of koala and wombat.

Because the *taki* inhabits a region that's easy to reach and work in, it provided the opportunity for an ecological study. No such study had been undertaken on a species of monkey-faced bat, and in 1992 I'd employed an honours student at Sydney University, Diana Fisher, to carry out the research under our Scott grant. She did her fieldwork between February and May, and her study remains a unique insight into these most unusual of creatures. It's tremendously time consuming to study bats in the vastness of a tropical forest, so Diana enlisted the help of local villagers. They cut tall bamboo poles and strung mist-nets on them in various parts of the forest, then watched all night for bats flying into the nets.

One of Diana's first discoveries was that, despite their

formidable teeth, the *taki* are gentle creatures. They almost never attempt to bite, even when being extracted from a mist-net, and they don't seem to be particularly scared of people. This gentle disposition is shared with many island species. Lacking predators, they don't anticipate harm from humans and so allow themselves to be handled without fuss.

The *taki* turned out to be reasonably abundant in the right habitat on two of the three large islands in the group, but were entirely absent on the third, Kolombangara, which had been devastated by logging in the 1970s. While Diana caught no *taki* there, she did meet people who recalled seeing them often prior to the logging operation. One man said that he saw the *taki* flying out from their roosts in panic as the last great trees crashed to earth. The bats disappeared into the distance, never to be seen again.

At first we found it difficult to understand why the *taki* should be so dependent on unlogged forest. After all, the fruits they eat abound in disturbed forests, including old gardens and village sites. The answer came when Diana fitted the bats with radio transmitters and began tracking them to their roosting sites. These were invariably enormous hollow trees growing in the fertile lowlands— the first targets of the loggers. Without roosts in which to rear their young and to keep safe by day the *taki* cannot persist.

Taki are quite social at their roost sites. Up to ten pack themselves like sardines into a single hollow a dozen metres or more up a great rainforest giant. They may do this because there are so few suitable roosting sites even in unlogged forest. During her work Diana saw another, larger species of bat sharing hollows with the *taki*. She never caught one, so could not identify it. To this day the creature's identity remains a great biological mystery. And there is an even greater mystery to be solved in the region, for villagers told Diana that they knew of giant rats in their forest. Given her focus on bats she was unable to pursue this lead, but it's almost certain that an unnamed species of giant rat lurks in the forests of Vangunu and New Georgia awaiting discovery by an adventurous biologist.

Towards the end of her stay Diana discovered something very surprising. Until 24 April, all of the *taki* she had examined had remained silent, even when being handled. She quite reasonably assumed that the species did not vocalise. But just after dusk on that night, a male she was holding in a canvas bag let out 'one very loud, high-pitched note'. This cry was immediately answered by a second animal, which was being held in a separate bag. From that night on the *taki* called to each other frequently—but only for the first hour or so after dusk.[20]

Just what the *taki* are saying to each other remains unknown, but we do have a few clues: sometimes one male appeared to be trying to drown out the call of a rival, while at other times females appeared to be answering males in a kind of duet. Among other mammals, such behaviours are common during the breeding season. But Diana observed that at the time the bats were calling most of the female *taki* were pregnant or lactating. It's possible that *taki* mate immediately after giving birth, but perhaps their vocal duels and duets have a different function. Whatever their meaning, the social lives of the *taki* were turning out to be more complex than we'd imagined.

It's a wonderful thing to glimpse the secret life of any species, and for us it was a rare pleasure indeed. Because of a lack of funding and time, all too often our work amounted to documenting the mere existence of a fauna whose ecology and social life remained absolutely unknown. But it was sad, too, to learn about the vulnerability of these gentle creatures to logging. As Diana was at work studying the *taki*, logging had already commenced on northern New Georgia and Vangunu. Nobody has returned to see how the bats are faring in the face of the relentless forest destruction, and little is being done to assist them. The report on this amazing creature that we wrote for the government of the Solomon Islands sits unread, I suspect, on the desk of some bureaucrat in Honiara.

4

FIJI AND NEW CALEDONIA

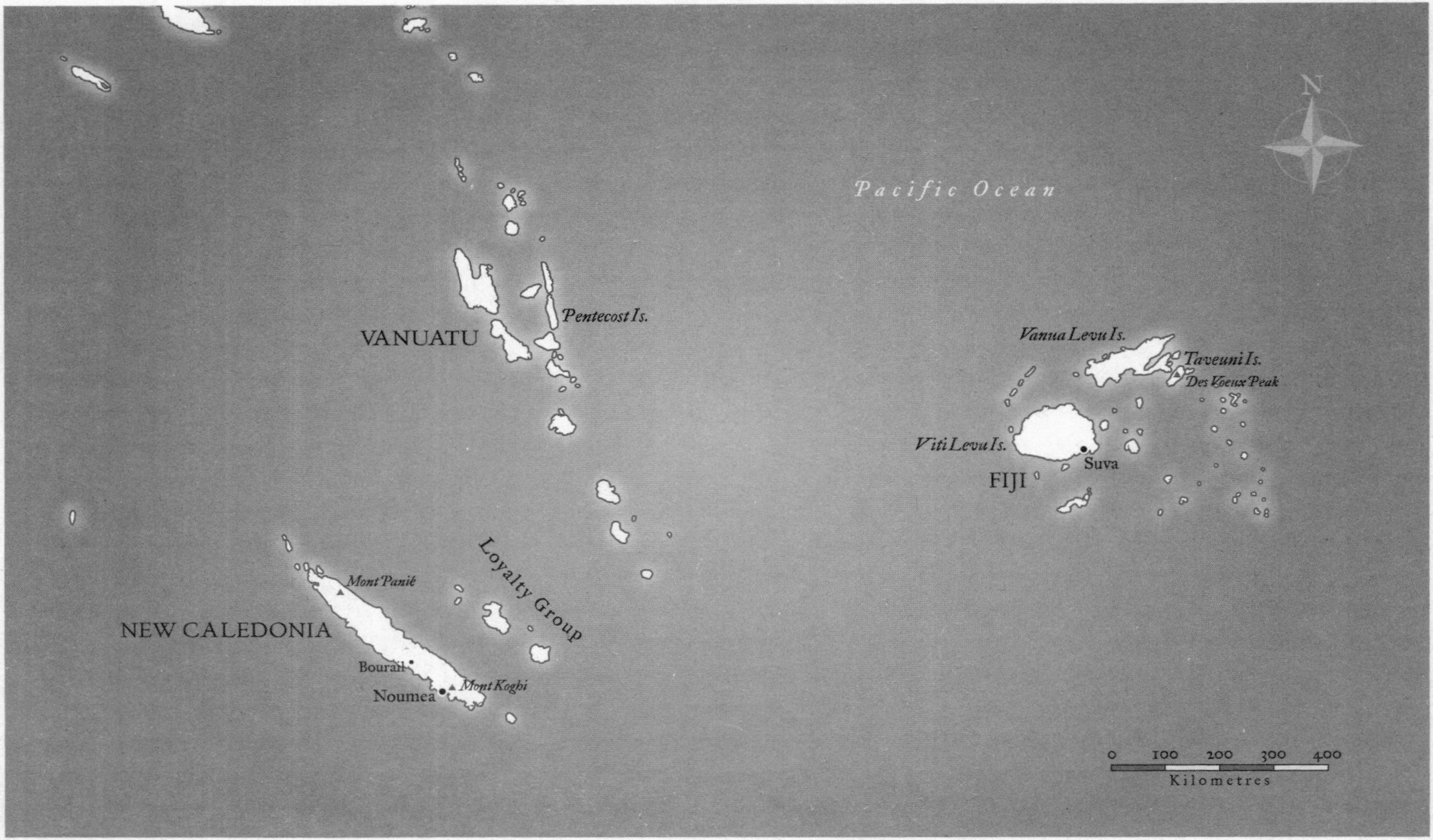
N
Pacific Ocean
VANUATU
Pentecost Is.
Vanua Levu Is.
Taveuni Is.
Des Voeux Peak
Viti Levu Is.
Suva
FIJI
Loyalty Group
Mont Panié
NEW CALEDONIA
Bourail
Mont Koghi
Noumea
0 100 200 300 400
Kilometres

A limit to our explorations had to be drawn somewhere amid the sprawling islands of the Pacific Ocean, and we decided that it should be at Fiji in the east and New Caledonia in the south. Beyond that, the size of the islands, their ecological complexity, and the number of mammal species each harbours drops rapidly away.

Fiji and New Caledonia are so distant from New Guinea that land mammals such as rats and marsupials have never reached them. But reptiles and birds have a greater capacity to cross the sea, and prior to human settlement both island groups were havens for bizarre, gigantic feathered and scaled creatures. The heaviest land creature on both island groups was a now-extinct horned tortoise. The shell of this gentle herbivore was around a metre long, and its tail was encased in bony armour that resembled a medieval mace. On its head were two horns resembling those of cows. On both island groups a similar top predator existed—a terrestrial crocodile, also now extinct, with a box-like snout and teeth that varied from sharp, cutting kinds in the front of the mouth to blunt crushing pegs at the back. At a metre or two from snout to tail, it was not fearsome to humans, but was clearly capable of consuming a range of smaller prey.

A variety of large, flightless birds also once inhabited these island archipelagoes. On Fiji a giant flightless pigeon, reminiscent of the dodo of Mauritius, was the largest of the feathered tribe, while on New Caledonia a giant megapode, a relative of Australia's brush turkey, filled the role. This creature would have been around a metre long, and it had a great bony knob atop its bill. Piles of stones found in various places around the island are interpreted as being the remains of the great mounds it accumulated in order to brood its eggs. The only native mammals on both island groups are bats. No extinct mammals have been described from either island group.

The inexorable tide of human settlement was slowed by the immensity of ocean separating Fiji and New Caledonia from their nearest neighbours, and it was only about 3000 years ago that humans first reached them. By then, the great pyramid of Giza was already an ancient monument. Though recent, the human impact was profound. Horned turtles, land crocodiles and gigantic flightless birds all disappeared, doubtless into that black hole that lies between nose and chin on the human anatomy. The pigs and dogs carried by the settlers must have done their bit, too, to hasten the demise of the island endemics, as did the rats that had hitch-hiked on the settlers' canoes. Like introduced rats elsewhere, they probably had a major, though as yet largely undocumented, impact on the smaller fauna.

The indigenous cultures of Fiji and New Caledonia differ from those of the Solomon Islands and most of New Guinea in that they are more hierarchical. Chiefs and men of status, who ruled over workers little better off than slaves, were a feature of traditional life in Fiji. By the time of the arrival of the Europeans political power seems to have been in the process of being manifested over very large areas. Indeed, by the early nineteenth century, on nearby Tonga, which had strong cultural links to Fiji, the leader Finau had consolidated power and established the region's first kingdom.

The Fiji group consists of more than three hundred separate islands, at the centre of which lies the large landmasses of Viti Levu and Vanua Levu. Both originated from volcanic activity and are tens of millions of years old. They have high mountains and are home to a great deal of unique biodiversity. The first European to sight the group was the Dutchman Abel Tasman, who passed by in 1643. Subsequent European mariners tended to avoid the place, deterred by tales of cannibalism and ferocious warriors. A few beachcombers, whalers, *bêche-de-mer* traders and assorted ne'er-do-wells, however, fitted into village life as best they could. In 1874 Fiji became a British colony, and sugar planters imported labourers from India to work the canefields. Today, the population of the islands is made up mostly of indigenous Fijians and the descendants of those Indians.

In contrast to Fiji, New Caledonia is an old continental fragment which had separated from eastern Australia around ninety million years ago. It has very few offshore islands. The Loyalty Islands (which are part of New Caledonia politically) are much more recent in origin and are biologically distinct. New Caledonia is renowned for its extraordinary biodiversity, particularly plants, the most ancient of which belong to lineages that last flourished globally during the age of the dinosaurs.

New Caledonia, along with Hawaii, was James Cook's greatest

discovery. Australia, New Zealand and other larger landmasses he charted had all been sighted earlier by European mariners. Cook encountered New Caledonia in 1774, during his second voyage, naming it New Caledonia (Caledonia being the name the Romans gave to northern Scotland) because the sparse vegetation of the island's nickel-rich mountains reminded him of heather-clad peaks. French missionaries began arriving in the 1840s, and in 1853 the French government formally annexed it. Like Australia, New Caledonia served as a penal colony.

Today, New Caledonia remains one of the world's last European colonies. Technically, it is known as a 'special collectivity' of France; only in July 2010 was it decided that the Kanak flag (the flag of the indigenous people) should fly beside the *tricoleur* as an official flag of the territory.

At the time we conducted our surveys the political environment of both Fiji and New Caledonia was changing quickly. Heading up the Fijian survey was Dr Sandra (Sandy) Ingleby, now collection manager in mammals at the Australian Museum, while I led the New Caledonian survey. Both surveys required multiple expeditions over a number of years, and Sandy was on occasion assisted by Pavel German and myself, while Alexandra Szalay assisted me on expeditions to New Caledonia.

CHAPTER 14

Bats at the Limit

In the nineteenth century Fiji had a bad reputation. Also known as the Cannibal Isles, it was a mariner's worst nightmare to be shipwrecked there. Yet many Europeans did visit, and some forged friendships with the locals. One such visitor was the New Englander William Endicott, who stayed in Bona-Ra-Ra village in 1831. His vessel the *Glide* was engaged in collecting *bêche-de-mer*—a sea-cucumber sold in China as an aphrodisiac. Because the *bêche-de-mer* has to undergo lengthy processing ashore, those involved in the business had to reach an accommodation with the islanders.

During Endicott's stay a raid on a mountain village had occurred, and the warriors of Bona-Ra-Ra brought back three bodies bound to long poles, each of which was carried ceremoniously into the village by six men. One body was given to a neighbouring village. One of the other bodies was recognised by an old woman as the killer of her son. Enraged, she took her revenge by filling a small dish with Kava and 'presenting it to the lips of the dead savage bade him drink...She then dashed the liquor in his face and broke the dish in pieces upon it.' Then, breaking a bamboo water container on his head, she told the men to dismember the body with the bamboo slivers. Endicott's account still makes vivid reading:

> They commenced...cutting up the body. The heads of both savages being now taken off, they next cut off the right hand and the left foot, right elbow and left knee, and so in like manner until all the limbs separated from the body.
>
> An oblong piece was then taken from the body commencing at the bottom of the chest and passing downwards about eight inches, and three or four inches wide at its broadest part. This was carefully laid aside for the King...The entrails and vitals were then taken out and cleansed for cooking... While this was going on, the *lobu* or oven was prepared... An excavation is made...[and] a large fire is then made in it, with small stones placed among the burning fuel...and as the bodies are cut to pieces they are thrown upon the fire, which after being thoroughly singed are scraped while hot by savages, who sit around the fire for this purpose. The skin by this process is made perfectly white...
>
> The head of the savage which was last taken off, was thrown towards the fire, and being thrown some distance it rolled a few feet from the men who were employed around it; when it was stolen by one of the savages who carried it behind the tree where I was sitting. He took the head in his lap and after combing away the hair from the top of it with his fingers picked out the pieces of the scull [sic] which was broken by the war club and commenced eating the brains...

I moved my position, the thief was discovered and was soon compelled to give up his booty.

The stones which had been placed upon the fire were now removed, the oven cleaned out, the flesh carefully and very neatly wrapped in fresh plantain leaves, and placed in it. The hot stones were also wrapped in leaves and placed among the flesh, and after it was all deposited in the oven, it was covered up two or three inches with the same kind of leaves, and the whole covered up with earth of sufficient depth to retain the heat.[21]

Cannibalism was so embedded in Fijian culture that the proper ritual greeting of a commoner to a chief was 'eat me'! Today, Fiji is an independent nation with a strong connection to its cultural traditions. My first exposure to the contradictions inherent in this history came while I was director of the South Australian Museum. I'd been asked to remove a picture of a Fijian cannibal feast from public display and was somewhat reluctant to comply, in part because the request came from a person who admitted knowing nothing about Fiji, and also because the picture was part of a gallery dating back to 1948—then the oldest intact museum exhibit in Australia. Some months later we received a visit from Fijian officials, and I asked their opinion on the matter. Their advice was to leave the picture in place. It was part of their heritage, they said, and they should not be ashamed of it.

Cannibalism sends such a shudder through us that we tend to grasp instinctively for moral, rather than biological explanations, for its existence. Yet even in our own society cannibalism has been accepted under certain circumstances, including the British navy, which implicitly condoned cannibalism among sailors who had resorted to it to survive. I think that biology, rather than ethics, offers the best explanations for both the taboo most of us observe against cannibalism, as well as the practice itself.

From an evolutionary perspective, the reason that such a strict taboo against cannibalism exists among most humans, most of the time, is it can incur high biological costs. These include the risk of parasite transmission as well the transmission of some very nasty diseases. Kuru is a brain disease (related to mad cow disease), which spreads only through cannibalism. It was prevalent among the women and children of the Fore people of New Guinea who, as part of the mortuary ritual, ate portions of relatives who had died. Prior to the outlawing of cannibalism by the Australian administration in the 1960s, the kuru epidemic was so severe that some villages were populated almost entirely by adult males.

But we must also remember that cannibalism can offer its participants benefits as well, in the form of a large return of protein. And in some circumstances, such as that faced by starving castaway sailors, that benefit can outweigh the costs. Among island populations also the cost-benefit ratio is altered. This is in part because island populations are often too small to sustain the diseases and parasites that are present on the mainland. Indeed the Fijian Ratu Udre Udre, who lived during the nineteenth century, purportedly ate hundreds of people without evident ill effect. Also islands often support few large land-based creatures, meaning that protein is at a premium for their human inhabitants. This suggests that if the benefits of cannibalism are likely to outweigh the costs anywhere, it's on islands. And indeed it's among the islands of the Pacific that cannibalism flourished.

Recent genetic studies have shown that the Fijians have a unique history. All Fijian men possess Y-chromosomes of Melanesian origin, while Fijian women have mitochondrial DNA of Polynesian type.[22] The most logical explanation for this is that, at the time the ancestors of the Polynesians were expanding through Melanesia, a canoe was hijacked by Melanesians and the males done away with. The hijackers voyaged into the Pacific, discovering the virginal islands of Fiji, which they then commenced peopling with their offspring from Polynesian women.

Fiji has a curious colonial history, which helped preserve traditional culture. Rather than do away with the *ratu* (chiefs), in 1876 the British administrator John Thurston established the Great Council of Chiefs, which was to act in an advisory role to the colonial government. Over a century later, in the 1980s, this was to have unfortunate consequences for democracy in Fiji.

If, arriving in Suva on the island of Viti Levu, you decide to stay at the Grand Pacific Hotel, you might feel as if the colonial era has never ended. The place is set by the water, with high-ceilinged rooms, lazy overhead fans and impeccably groomed waiting staff of the kind that you read about in Somerset Maugham novels. For a few days, as Sandy and I organised permits, travel and equipment, we enjoyed the atmosphere as well as a few evening gin-and-tonics there. But soon the day came for us to head into the hills—we'd decided that we should begin by surveying Viti Levu's highest peaks. Our base would be a village near the Monasavu dam, at an elevation of around 1000 metres. At this altitude the forest is little more than alpine scrub, but the species present are intriguing.

One of the most interesting plants to occur there is the beautiful whale-tooth pine (*Acmopyle sahniana*). Its Fijian name *draubata* comes from its leaves, which are dark green on top and white below and resembling in shape the sperm-whale teeth that are the most esteemed possessions of traditional Fijians. The plant is an ancient relic, known only from two tiny populations that cling to razor-edge ridgetops in the highest mountains of Viti Levu. Fifty million years ago, however, its relatives were widespread in southeast Australia including Tasmania. For reasons that remain unknown, by thirty-five million years ago the genus had disappeared everywhere except New Caledonia and Fiji.

Before we could explore this amazing natural world, we had to seek the permission of the local village chief, and that, we soon learned, meant making a gift. A decision on the type of gift required was quickly made, for we heard through a friend that the chief would like a dinner set of English china—a twelve piece setting, of a quality suitable for entertaining other chiefs. I gulped somewhat at the expenditure and, hoping that the spirit of Miss Scott would understand, went out and bought the best set I could find in Suva.

But there was one more hurdle. The men of the expedition—namely me—would have to drink kava with the chief. Kava is an intoxicant made from the roots of a bush-like member of the pepper family that is native to the Pacific Islands. Its stems look rather like fleshy, knobbly canes. Great bunches of the plant with roots attached can be seen for sale in markets and by roadsides throughout the islands. I had drunk kava once before, in Vanuatu, when I was invited to the *nakamal* (kava-drinking hut) patronised by Father Walter Lini, who was then prime minister of Vanuatu.

The *nakamal* was a small wooden structure with a dirt floor, and bench-seats and counter made of branches round the walls. The kava was served in a half-coconut shell and it was, I recall, ghastly looking—reminiscent of very dirty dishwater, and tasting little better. I managed to force down a few cups, after which I felt supremely calm—indeed almost cosmically omniscient. But when I tried to stand up I found that my legs were like rubber bands. I had almost no control over my body, and writing, I later learned, was entirely beyond me.

As my eyes adjusted to the gloom of the hut I saw that the long-term effects of drinking kava were even worse. Old men, their skin flaking from their bodies, their muscles slack and their eyes red and staring, sat on the benches around me. They looked like something out of an old horror film, but they were clearly having a good time. Several were from Pentecost—the island where young men practise land diving, from which the modern sport of bungee-jumping is derived. The young men of Pentecost are required to jump from high platforms with vines tied to their ankles that pull them up just as they kiss the mud below. It takes some courage, and one of my drinking companions was adamant that he had the courage—however his unconquerable fear of heights precluded him from participating!

In Vanuatu, the kava drink is made by mincing the roots of the kava plant in a meat grinder and mixing the pulp with water. There is little ceremony involved in drinking it at a *nakamal*, other than a mighty gulp accompanied by a clap of hands and swallowing a biscuit or a piece of dried fish. But things are very different in Fiji. There, traditionally, the drink had been made by young women who chewed the roots and spat the pulp out into a large, ornate and highly prized kava bowl. From there it was doled out into half-coconut shells, and consumed with a protocol as intricate as a Japanese tea ceremony. As it happened, we were going to one of the most traditional villages in Fiji—I wondered how much of the kava tradition survived there.

Quite aside from my misgivings about the local manufacturing process, I was worried about the impact of the kava on my body. How could I possibly work under its influence? When we arrived at the village the chief was waiting for us. He accepted our gift of the dinner setting, then ushered us into an open-sided hut where half the village seemed to be assembled on mats on the floor. Near the chief lay the kava bowl, glazed from innumerable ceremonies and already full of the greyish liquid. A circle of drinkers had formed, and I was bade to sit cross-legged on a mat near the chief while he made a speech in Fijian. So awed was I by the formality that I dared not ask how the kava was made, or what the chief had said.

When the coconut shell was offered to me, I dutifully emptied it at a gulp. Relieved, and thinking the ceremony over, I made to unbend my legs, but stopped when the chief said something and the crowd burst into laughter. Apparently we had to empty the contents of the whole kava bowl and to cut and run now was considered impolite. Somehow I got through four or five more shells-full. When I got to my feet, miraculously I found I could stand and walk without feeling like I was on the moon. Fijian kava, I found out, is far less intoxicating that the Vanuatu variety. A Fijian friend also told me what the chief had said when I went to get up the first time: 'Hey, he's leaving because he's scared he'll piss the bed tonight from drinking too much kava!'

Since gaining its independence from Britain in 1970, Fiji's political scene has been rocked by military coups, and I later learned that the chief I was drinking with was one of those behind the first coup, in 1987. The Great Council of Chiefs met once every year or two in a splendid ceremony, but as democracy began to be implemented in the 1960s the council's influence waned, and upon independence its only power was to appoint eight of Fiji's twenty-two senators. Perhaps disgruntled by the loss of influence, its members backed Sitiveni Rabuka in a military coup. The Great

Council was rewarded by having the senate turned into an aristo-cratic body peopled solely by chiefs. But having dabbled in politics, the Great Council has since paid a high price. It was suspended in 2007 by Fiji's first dictator, Frank Bainimarama. Its remaining authority had been sullied by its association with power. Today Fiji is in the grip of a full military dictatorship—the first in the Pacific region.

After leaving the kava hut we set to work stringing our mist-nets. The vegetation looked so low and scrubby that my expectations too were low. The following morning, however, we found several very unusual fruit bats in the mesh. They were a dark olive colour and where they were not naked they were covered in very short fur. But it was their long, powerful legs, and long bony tail—rather like that of a rat—that struck me. Indeed, overall they looked rather shrew-like, as if they were a living relic of the first bats to make the transition to flying. They were Fijian blossom bats, and DNA studies undertaken with the help of the samples we collected subsequently demonstrated that they are indeed very primitive members of the fruit-bat family. Their ancestors had presumably reached Fiji many millions of years ago when the fruit bats were only beginning to diversify. Elsewhere, such primitive species have either become extinct or evolved into something else. Fiji provides a refuge for this living fossil.

We remained on Viti Levu for about a week, surveying the Monasavu Plateau, but found no other bat species of particular interest. I would return to Fiji, however, with Pavel German, some years later, in order to survey another island in the Fijian archi-pelago.

CHAPTER 15

Fiji's Garden Isle

Pavel German is one of the most capable self-taught field biologists and photographers I've ever met. Born in Russia, he had worked for many years as a biological collector for the state. This often involved travel to remote regions of the Soviet empire, such as the Muslim south, where infrastructure was rudimentary and life cheap. He was even granted permission to enter the forests around secret military bases and industrial sites, and had seen more of the Soviet Union than almost anybody else. Travelling in such regions is perilous enough, but carrying huge vats of preserving alcohol (which was often more useful than cash as a currency), Pavel was a prime target for robbers. He may well be alive today only because he's a first-class boxer as well as a highly intelligent and perceptive person.

In the 1980s Pavel fled the Soviet Union, and eventually came to Australia where he began work at Taronga Zoo in Sydney. He had given this up, however, to pursue a career in photography at the time I began planning the Scott expeditions. In addition to his experience of working in difficult circumstances, his photographic skills were of enormous value to us. Indeed, Pavel remains the only person to have photographed some of Melanesia's rarest and most obscure fauna.

After his experiences in the Soviet Union, Pavel found work in Melanesia a breeze. Even where his language skills were limited he had a way of establishing a rapport with people, and everywhere he went the locals loved him. After a hot walk, he would astonish them by drinking the contents of three coconuts in rapid succession, and the kids would vie with each other to climb the palms to chop more, eager to see just how many nuts this strange visitor could dispose of. Of an evening he'd tell stories about Australia or Russia, then amaze the village by beating all of its young men at arm wrestling. Even his loud snoring, which was my only frustration when travelling with him, was a source of fascination to the villagers.

Taveuni is Fiji's garden island. Covered in lush green forest, it is the third-largest island in the group, and has escaped such scourges as the introduced mongoose. As a result, it is the only island in the Fijis to have retained all six bat species native to the group. In 1990 Pavel and I travelled to the island to investigate the biodiversity of Des Voeux Peak, the second-highest mountain on the island. Its unique biodiversity is legendary. Among the species the peak is renowned for is a creeper known to Fijians as *tagimaucia* (*Medinilla waterhousei*), which has been described as 'Fiji's pride and joy'.[23] Its brilliant scarlet-and-white blooms stand out like burning beacons in the ocean of green that swathes the peak, and can be seen from quite a distance. Des Voeux Peak is also home to one of Fiji's most unusual birds—the silktail. This

tiny iridescent creature is placed in its own family. It had clearly
evolved in isolation for a very long time, suggesting that its ancestors
made their way to Fiji many millions of years ago. It was once
believed to be a distant relative of the birds of paradise, but is now
thought to be related to the monarch flycatchers. The species is only
found on Taveuni and the adjacent island of Vanua Levu, and it
is most easily seen on Des Voeux Peak. I was keen to see this
living jewel.

Arriving by air, we soon settled into Somosomo village. It had
a basic guesthouse and was located on the road leading to the peak,
and so formed a good base for our explorations. The village was the
ancestral home of the powerful kings of Somosomo, and it had a
fabulous history. As the seat of power for much of Taveuni and the
surrounding area, it had a reputation for savagery that chilled the
blood of Fijians throughout the archipelago. As one early missionary wrote, 'Even in the other islands Somosomo was spoken of as a
place of dreadful cannibalism.'[24] Yet, astonishingly, Somosomo was
destined to become the site of one of the earliest mission stations in
the Fiji islands.

The Wesleyan missionaries who arrived there in 1839 had been
encouraged to come by the king's son. They should have suspected
that his motives did not stem entirely from a yearning for salvation—for he told them that as 'muskets and gunpowder are true,
[so] your religion must be true'.[25] According to James Calvert, the
chronicler of Fiji's early missionaries, once the Wesleyans were
settled at Somosomo with their families, they found themselves
surrounded by 'all the horrors of Fijian life in an unmixed and
unmodified form'.[26]

The mission's establishment could not have got off to a worse
start. Within weeks of the arrival of the missionaries the favourite

son of King Tuithaku was shipwrecked and then eaten by a hostile group. Suspicions arose in the village that this horrid misfortune had somehow been visited on the royal household as a result of their acceptance of the missionaries, and relations at once cooled. But worse, tradition demanded that several of the eaten prince's wives be strangled, so that they could accompany him in the afterlife. When they learned of this, the missionaries pleaded with the grieving king for the women to be spared, but in anger at having his authority challenged the king increased the number to be strangled to sixteen, and had the whole lot buried just outside the door to the missionaries' house.

Had I been a Wesleyan missionary, I think I might have left at this point. But they were clearly made of sterner stuff than I, and they stood their ground, virtually blockading themselves and their families into the thatch hut that was their home. Unfortunately for them, the village's cannibal cooking ovens were situated just outside one of their windows, and the constant scenes of butchery and the fumes arising from the pit forced them to keep their blinds closed, depriving them of light and air. One can only imagine the conversations that must have gone on between husbands, wives and children in the stifling heat of the darkened hut beset on all sides by what were in their eyes scenes of utmost depravity. The great question must surely have been, 'will we be next?'

Things did improve briefly for the doughty missionaries when a young chief became ill and was cured by a pastor. The chief was a strapping lad, 'a head and neck taller than myself, and three times the bulk, every part indicating the strength of a giant', according to a European who measured him. But even the chief's gratitude caused discomfort, for when he turned up naked at the mission to give thanks for his recovery, the sight of him 'was enough to frighten Mrs Brooke'—a fright which seems to have turned into hysteria when the naked giant took her seven-week-old infant son into his arms and 'put his great tongue in his mouth'. By 1847 the missionaries had fled to less challenging fields, leaving Somosomo to its traditional ways.

A detailed insight into life in traditional Somosomo was published by Thomas Williams, a missionary who visited the besieged mission station around the middle of 1845. Old King Tuithaku was by then feeble, and Williams went to visit him on several occasions. Knowing something of Fijian funerary rites, he was delighted to find the king's health steadily improving. He was mightily surprised on 24 August, therefore, at being told that the king was dead and that preparations were being made, according to custom, to strangle all of his wives. Williams rushed to the royal house, determined to save the women, only to find that the killings were already under way. He wrote:

> The effect of the scene was overwhelming. Scores of deliberate murderers, in the very act, surrounded me: yet there was no confusion, and, except a word from him who presided, no noise, but only an unearthly, horrid stillness...My arrival was during a hush, just at the crisis of death...
>
> Occupying the centre of that large room were two groups, the business of which could not be mistaken. All sat on the floor; the middle figure of each group being held in a sitting posture by several females, and hidden by a large veil. On either side of each veiled figure was a company of eight or ten strong men, one company hauling against the other on a white cord, which was passed twice round the neck of the doomed one, who thus, in a few minutes, ceased to live.

As my self-command was returning, the group furthest from me began to move; the men slackened their hold, and the attendant women removed the large covering, making it into a couch for the victim. As that veil was lifted, some of the men beheld the distorted features of a mother, whom they had helped to murder, and smiled with satisfaction as the corpse was laid out for decoration. Convulsive struggles on the part of the poor creature near me showed that she still lived. She was a stout woman, and some of the executioners jocosely invited those who sat near to have pity, and help them. At length the women said, 'She is cold.' The fatal cord fell; and, as the covering was raised, I saw dead the obedient wife and unwearied attendant of the old King. Leaving the women to adjust her hair, oil her body, cover the face with vermillion and adorn her with flowers, I passed on to see the remains of the deceased Tuithaku.[27]

When he arrived at the royal death bed, however, Williams was astonished to find that the old man, though somewhat feeble, was clearly still alive and kicking! Perplexed, he approached the king's son who, 'seemed greatly moved, put his arm round and embraced me, saying, before I could speak. "See, the father of us two is dead...his spirit is gone. You see his body move; but that it does unconsciously."'[28]

Williams was now faced with a cruel dilemma. Should he try to save the remaining women and allow the old king to be buried alive, or should he argue that the king was in fact not dead, and risk the possibility that he would die later and the women be strangled while Williams was absent? Reluctantly, Williams decided not to debate the state of the king's health, instead requesting that the remaining women be spared. Out of respect for the visiting

European the request was granted, but not everyone was happy. 'Why is it that I am not to be strangled?' wailed one wife, who was consoled with the fabricated news that there was nobody present of sufficiently high rank to do the deed for her.[29]

After the role he had played in the affair, Williams could not bring himself to attend the king's burial, but he was told by an observer that the king was heard to cough 'after a considerable quantity of soil had been thrown in the grave'.[30]

As I sat in the lonely guesthouse in Somosomo, reading these tales of Taveuni's past to Pavel by the light of a kerosene lamp, my skin tingled, and later that night my dreams were disturbed by gory scenes. The cannibal feasts, strangling and live burial had all happened right here in Somosomo, the place where we had just enjoyed a cold beer and a meal. Perhaps, I mused, the royal hut was located nearby; we were after all near the very centre of Somosomo. One hundred and fifty years—two human lifetimes—separated those days from mine. Yet the scenes Williams described seemed somehow hauntingly present.

Pavel and I had less than a week together in Taveuni before I went on to New Caledonia for more survey work. A paved road leads to the summit of Des Voeux Peak, which at 1190 metres is the second-highest point on Taveuni. The road, which serves a telecommunications tower, was a rare luxury in Melanesia. With our hired car it took less than an hour to reach the summit, and as we drove the road each morning and evening we occasionally saw scenes reminiscent of an earlier era. One morning we encountered a Fijian on a horse. He could have been a facsimile of the giant who stuck his tongue into the mouth of Mrs Brooke's infant son, and he sat as proudly on his mount as a knight of old. He was, he said, going pig hunting, and in his hand was a lance of wood and iron that would have done a knight templar proud.

By the end of our first day we had strung ten mist-nets at elevations between 880 metres and the summit. The object of our search was yet another monkey-faced bat. This species, the Fijian monkey-faced bat, had been described in 1978, and is the only species of monkey-face to occur outside the Solomon Islands. It is also the only mammal unique to Fiji. It was known from just two specimens. We were keen to see if we could learn something of its ecology, evolution, distribution and abundance. Des Voeux Peak receives the highest rainfall in Fiji. It's an exceptionally misty place—parts of it seem to be eternally swathed in cloud. Because of the persistent cloud above about 900 metres, a unique vegetation community has established itself. The tallest plants there are palms, ginger bushes and herbs, and delicate lichens, mosses and diaphanous ferns abound. It looked like a kind of fairyland, untouched by the outside world.

This illusion of virginal nature was shattered, however, when we reached the summit itself. When we returned that evening we found that spotlights kept the telecommunications tower constantly illuminated, and that they were attracting innumerable insects. Somehow cane toads had arrived even in this remote spot, and the ground below the lights seethed with huge, ugly specimens.

Our few days on the mountain had yielded no sign of the Fijian monkey-faced bat. We had, however, netted a few Fijian flying foxes, which were of mild interest. On my last morning on Taveuni it was freezing and we checked the mist-nets in the rain, only to find them once again holding just a couple of common flying foxes. My heart sank. As Pavel drove me to the airport various dismal possibilities occupied my mind. Perhaps the Fijian monkey-face was already extinct, or so rare as to be effectively so. After all, its habitat was tiny—just the few tens of square kilometres that make up the highlands of Taveuni. There was, however, one remaining hope. Pavel had agreed to stay on in Somosomo for a few days, continuing the program.

To my amazement and delight, after I left he netted not one but three Fijian monkey-faced bats. A change in the weather may have been responsible for this upturn, for a series of misty nights occurred on the peak—conditions that evidently favour this rarest of bats. Pavel recorded that all of the monkey-faces were trapped in mist-nets set at elevations above 1000 metres, indicating that the species was indeed restricted to the mountain summits. This good fortune allowed Pavel to take a magnificent series of photographs, the first ever of this species. Like other monkey-faced bats, its wings meet in the midline of the back, giving it tremendous lift and, most likely, the ability to fly backwards. In its tangled, misty habitat such traits would be invaluable. Indeed, we hypothesised that it might be protected from competition with the more abundant Fijian flying foxes by the mist that wreaths the peaks, for we never caught Fijian flying foxes in nets we set on misty nights.

The records Pavel made hinted at an unusual biology. The female monkey-faces had tiny teats, and when they were touched the milk shot from them with force. Most flying foxes, including other monkey-faces, have much larger nipples, which their newborns cling onto as they fly. Some bats even have false nipples in their groin to which the young attach themselves, providing more freedom to their mothers to manoeuvre their wings. The nipples of the Fijian monkey-face mothers were far too small to carry young, so we reasoned that they must keep their newborns safely hidden, perhaps in a tree-hollow, while they forage. Just how the young could stay warm on chilly Des Voeux Peak, however, we could not imagine.

Pavel's photographs reveal a bat with ears so short that they're entirely hidden in the silver-tipped fur of its head, eyes that shine like bright orange jewels and a very solid snout. The male and female differ in colour: the back of the female is khaki, while that of the male is golden. Colour differences between the sexes are rare among bats. Combined with what we know of its reproduction and what is known about the behaviour of the *taki*, the colour difference suggests that the monkey-faced bats can have a rich and unusual social life, but its nature remains almost entirely unknown.

As a result of Pavel's work, the Fijian monkey-faced bat is now recognised by the IUCN as critically endangered. Further studies are under way and it is hoped they will lead to effective conservation. Of all the threats it faces, climate change is doubtless the most severe, for all mountaintops world-wide are warming, and the cold-adapted ecosystems at their summits are shrinking. Such a threat is difficult to counter, and the best that can be done for now is probably continued monitoring of the species.

Pavel's work yielded one more important finding. DNA samples he collected revealed that the Fijian monkey-faced bat is only distantly related to the monkey-faces of the Solomon Islands. Indeed, they may have evolved the characteristics they share— such as wings that join at the midline, complex teeth and robust muzzle—independently. If so, this would be a striking case of convergent evolution—rather akin to that of the evolutions of the wolf and the thylacine. This discovery has prompted Kris Helgen to re-classify the Fijian monkey-face by taking it out of the genus *Pteralopex* (winged Arctic fox) and creating a new genus name for it—*Mirimiri*, the Fijian word for mist.

CHAPTER 16

Nouvelle Calédonie

From Taveuni I flew to Suva, then on to Tontouta airport in New Caledonia. The island has always intrigued me. It is an ancient fragment of the continent of Gondwana which separated from the east coast of Australia ninety million years ago, and it has remained isolated ever since. Many species found there trace their ancestry back to this distant time, making it an ark filled with life forms that last flourished elsewhere during the age of the dinosaurs. Of most significance to our work, however, only eight native mammals were known from the island, which is a small number given the island's large size and age. One possible reason for this was that New Caledonia might have its own as yet undiscovered species of monkey-faced bat. Hard work and much mountain climbing might just reveal its presence.

I was joined in much of the survey work by Alexandra Szalay, who had worked in the Australian Museum's anthropology department prior to joining mammalogy. Her knowledge of Melanesian cultures and their use of natural resources was invaluable, as was her curatorial expertise, which saw the specimens we collected promptly and accurately catalogued. Long after the fieldwork documented here she was to become my wife.

We were unaware that we had arrived in New Caledonia at the commencement of the Feast of the Assumption, and were concerned to find that the Caldoche, as islanders of French descent are called, celebrate it as a serious and lengthy holiday. This made it impossible to meet the officials we needed to see in Noumea, and we realised that we must stay in town—which was both expensive and frustrating—until the government offices re-opened. Armed with a collecting permit, obtained in advance, we could, however, investigate localities within reach of the town. The most interesting of these was Mont Koghi, a forest-covered mountain that lay almost on Noumea's suburban margins.

Working in New Caledonia can be very a different experience from working elsewhere in Melanesia. Paved roads are common and good food is easy to find. Nonetheless, we were bemused to learn that a paved road led right to the forest edge on Mont Koghi, and that just a hundred metres away was a café that served *petits fours* and excellent coffee. We set the mist-nets in spitting rain, and as the light faded retired to the café. For once it felt that the mammalogists had it as easy as the ornithologists.

I hardly expected to find anything interesting at such an accessible location. The café closed and, as darkness fell, we strolled to the mist-nets. Two small bats were hanging in them, their fur glistening with tiny droplets of mist-like rain. Their ears were enormously long and pleated, and on the end of their nose a short, fleshy leaf-like structure was prominent. This made identification easy. They were Australasian long-eared bats—a genus never before recorded from New Caledonia. Thrilled at the ease of this major discovery, we placed the bats in canvas bags, took down the nets and prepared to return to Noumea for a celebratory drink.

The Australasian long-eared bats are notoriously difficult to classify to a species, but fortuitously the world authority was working on the Scott grant. Dr Harry Parnaby had written his doctoral thesis on the group, and he would go on to identify the New Caledonian specimens as a new species, which he named *Nyctophilus nebulosus*. The name refers to the rather nebulous physical differences that characterise the various kinds of long-eared bats, which genetic studies nonetheless tell us are distinct species. It was, Harry concluded, a close relative of an Australian species whose ancestors must have been blown to New Caledonia, probably within the last million years. Prior to this discovery the whole mammal fauna of New Caledonia had consisted of just eight species. To add a ninth was a satisfying achievement.

As we drove back down the road towards Noumea Alex spotted what looked like a small dinosaur. The size of my hand, it stood tall on its legs, and appeared to have two blunt horns at the base of its head. It was, I later found out, a knob-headed giant gecko, and it didn't seem to mind at all when we pulled over and picked it up. Only much later I learned that the species can inflict a nasty bite and is quite capable of killing and eating other lizards. But this one was gentle in the hand as I examined and photographed it.

New Caledonia's giant geckoes are many and varied, and among them is the world's largest surviving gecko species— Leache's giant gecko—which can grow to the length of a man's forearm and almost as thick. Giant geckoes have been present on New Caledonia for a very long time—perhaps ever since the landmass broke away from Australia. And on an island with few competing species they've evolved to take advantage of many ecological niches. The largest species eat fruit, and thus take the place of possums and monkeys elsewhere. Smaller but still impressive species like Jean-Claude Gecko—as Alex named our new friend—replace carnivores such as stoats and quolls elsewhere. Such species are known as ecological vicars because, like vicars in the Anglican church, they stand in for somebody else.

As the holiday marking the Feast of the Assumption dragged on, we realised that it would be some time before anybody returned to work, so we looked about for ways to fill in our time. High on the priority list was a visit to Parc Forestier which, in addition to having a botanic garden filled with unique New Caledonian flora and in which all the trees are identified, has a small zoo. It was awe-inspiring to wander through the groves of ancient pines and other plants that had become extinct elsewhere millions of years ago, and I yearned for the opportunity to see them growing in natural conditions.

The zoo's main exhibits are the island's unique fauna, including the intriguing kagu. A large, pale-grey forest bird, it's placed in its own family, its nearest relatives being the sun-bittern of South America and the extinct moa-sized adze-bill of New Zealand. It's an example of a species that arrived in its island home by going the long way round—arriving across the wide Pacific from South America. Very few species have achieved this, though the ancestors of New Zealand's short-tailed bats and Fiji's iguanas must have made a similar journey. These voyagers against the wind are ancient migrants indeed, having arrived tens of millions of years ago. The kagu and short-tailed bats have been on their islands so long they're classified in their own families. Just why the American immigrants are so few and ancient is not clear, but it's possible that in earlier times winds or ocean currents made the long Pacific crossing from South America easier than it is today.

I longed to see a kagu in the wild, but their rarity precluded that privilege. I watched the pair in the Parc Forestier with great interest. The eyes of these all-but-flightless, carnivorous birds are very large, and their red bill is stout. But it's their ridiculously floppy crest and bright patches of feathers, hidden in the folded wings, that give them much of their character. Surely, these feathers must be used for display—perhaps to a potential mate or a competitor.

Aspects of the kagu's biology remain mysterious. Why should it, alone among birds, have one third of the usual number of red blood cells while those cells are three times as capable of carrying oxygen? And why does it have peculiar nasal corns—again unique among birds—covering its nostrils? The kagu is now so rare that it is difficult to study, but its breeding biology reveals a bit about why it has fared so poorly in the face of the European invasion. Kagus are monogamous. The female lays a single egg on the forest floor, which she makes no effort at all to conceal. Even though the offspring of previous years will stay around the nest to help raise and protect the young, a kagu egg or chick must be an obvious and irresistible delicacy for a pig or rat. And so it is that the kagu finds itself on the endangered list.

Fired up by our success on Mont Koghi, and with the Feast of the Assumption at last behind us, Alex and I set off to see what the further reaches of this great island might offer. A colleague in the forestry department had suggested that we stay at a forest camp in a place called Col d'Amieu. Located at an elevation of around 400 metres in the middle of the island, it suited us well as a base from which to explore the mid-elevation forests.

After the luxury of a hotel in Noumea, accommodations at the forestry station were rudimentary—sawdust-filled mattresses on the floor in a cabin made of rough-hewn timber, an outside privy and no cooking facilities beyond our Trangia. But it was set in an idyllic forest, and the air was crisp and clean. After setting our mist-nets, which due to the terrain was more exhausting than it was at Mont Koghi, we retired at dusk to the cabin, and an early night's sleep.

Long before dawn we woke to the jingle of metal, the creak of leather and low, equine snorting. I was unsure whether I was awake or dreaming; it was only the smell of frying sausages and eggs and the aroma of coffee brewing in the pre-dawn gloom that convinced me we had visitors. Cautiously, I opened the door to find a troop of gendarmes camped under the eaves. Dressed in handsome, if somewhat soiled, grey uniforms, they were a cavalry regiment—their horses stood calmly nearby. As they gathered round a small fire in the chilly pre-dawn light they looked like a troop of Confederate soldiers. Yet all except the officer were black-skinned Kanaks—the original Melanesian inhabitants of New Caledonia.

When I greeted them they explained that they were on patrol for rebels. At the time of our visit New Caledonia was still under colonial control, and to many Kanaks the colony seemed to be moving towards independence at glacial speed. Kanaks still make up just under half of the population and, prior to the Matignon Agreement, which was brokered in 1988, some were in open revolt.

The bloody events of the 1980s were only the final spasms of a conflict that had begun with French colonisation. In 1878 the Kanaks had attempted to drive the French from the island. Around a thousand were killed in the conflict, but that did not prevent a further revolt in 1917 or the establishment of an independence

movement in the 1970s. By late 1984 the rebels had formed a liberation army and set up a provisional government. In response, the
Caldoche ambushed and killed ten Kanak leaders. In May 1988
the violence again reached flashpoint, with Kanaks holding French
gendarmes hostage in a cave on the island of Ouvéa in the Loyalty
Group. Nineteen Kanak liberation leaders and fighters were killed
in the rescue attempt—some while in French custody. Two years
later, the search for rebels holding out in New Caledonia's rugged
central ranges clearly remained unfinished.

I hoped that our research in New Caledonia might yield
a monkey-faced bat. After all, the island has many high peaks,
and is not all that distant from the Solomons and Fiji. Instead, at
Col d'Amieu, our nets trapped a diminutive flying fox which, until
the 1960s, had been known from just two specimens. The New
Caledonia flying fox is a diminutive, charcoal-coloured creature
which like the monkey-faced bats has tiny ears, almost hidden
in its fur, and multi-cusped teeth. Genetic studies reveal that the
New Caledonia flying fox is not a particularly close relative of the
monkey-faces, but perhaps it illustrates a stage in the evolution of
the group, a time when it was just branching off from its ancestors
and acquiring the features that make it so distinctive.

As the holiday had prevented me from obtaining formaldehyde
in which to preserve specimens, I needed to obtain a supply. North
of Noumea is the small town of Bourail, and I was relieved to find
its pharmacy was open. The only difficulty I faced in obtaining
the formaldehyde was my lack of French. Not to be put off by
such trivial things, I approached the counter, behind which stood
a rather serious-looking lady pharmacist in a white coat, and said,
'Avez vous le préservatif?' Préservatif is one of those French words
which seems to be almost deliberately designed to confuse an

anglophone zoologist, who uses preservatives to preserve his animal specimens. In fact *un préservatif* is a condom.

The pharmacist fumbled about behind the counter, and eventually re-emerged holding a small brown paper bag. Thinking that she'd tried to palm me off with just a few millilitres of formaldehyde, when I needed at least half a litre, I said in my appallingly accented French, '*Non, plus grand. Un demi-litre.*' At this my pharmacist flushed red, and simply stood there. Hopelessly confused by this stalemate, and suspecting by this time that there might be more than one kind of *préservatif,* I tried to clarify matters by asking, carefully, '*Avez vous le préservatif pour les animaux morts?*'

My ignorance of French was such that I didn't know that I'd just asked for condoms for dead animals, so the chemist's explosion caught me unawares. The small paper bag flew in the air as she raised her hands above her head in an effort to drive what was clearly a pervert from her pharmacy. By this time, a small crowd of spectators had assembled, amongst whom was an elderly man. He stepped forward, asking in broken English what it was in fact that I wanted. '*Ah, le formol,*' he said with a kind smile, and soon I had a small bottle of the precious *préservatif* in my hands.

One of my worst character flaws is not knowing my limits. We had many winding roads to drive, so I felt I should buy some travel-sickness pills. Again mustering my best French, I went to the traumatised lady behind the counter and said, rather grandly, '*Avez vous le medecin pour le mal de travail?*' There was something about her response to my request for 'work sickness pills' that had me realise that I would just have to put up with the discomfort.

Our next destination was Mont Dzumac, a desolate-looking peak that is one of the highest in the south of the island. Driving the winding dirt road to the summit was challenging, but fascinating, for the *maquis*, as the scrubby heath flora of New Caledonia is known, was in full flower. Magnificent red bells, six centimetres or more long, hung from low tufted plants, and shrubs were decked in spectacular flowers of a variety of colours. Like Australia's heathlands, this vegetation owes much of its character to the soil it grows in. The rocks of New Caledonia were formed deep in the Earth's crust, and they contain high concentrations of metals such as nickel. These are toxic to many plants, and only a few stunted varieties can thrive there. Yet these species must attract pollinators, which are few: hence the extravagant flowers.

Prior to leaving Australia I had met a Caldoche named Jean-Pierre Revercé. He was the deputy mayor of Bourail, and he'd invited us to visit if ever we were in the area. When we met up he took us to his seaside shack, which stood alone just behind the beach in a remote cove. It was surrounded by a forest of native trees covered in brilliant white gardenia-like flowers. Later, I learned that these were true members of the gardenia family and that they were native to the sandy coastal regions of the island. The shack was raised on poles and was largely without walls. Looking out over the bush to the turquoise sea, it was easy to imagine ourselves in paradise.

That afternoon we went out fishing in the lagoon, and I got to know Jean-Pierre a little better. 'Do you want me to put your hook where the fish are?' he asked, before tossing my line for me in a curious way. After some time without a bite I noticed Jean-Pierre quietly chuckling and, intrigued, I decided to trace my line. The hook was in our bait box, which was indeed full of bait fish! Despite such joking we caught enough fish for an abundant supper, but Jean-Pierre was not yet satisfied. 'We must 'ave mud-crabs,' he said, and set out at last light with a torch and pole. True to his word, he returned in an hour with an enormous crab, which joined the fish on the barbecue. But he had brought something else with him as well—a perfect New Caledonian nautilus shell, which the tide had deposited on the sand right next to his boat. The New Caledonian nautilus is perhaps the most beautiful of an exquisitely beautiful genus of shells—wave patterned in white and chocolate brown, its spiral forming a perfect Fibonacci sequence. He gave it to Alex with a flourish, saying, 'it must be a gift from the sea to you.'

CHAPTER 17

The Mystery of Mont Panié

All our searches thus far had failed to find a monkey-faced bat on New Caledonia, and there was only one realistic option remaining: Mont Panié. At just over 1600 metres high it's the tallest peak on New Caledonia, and on its upper slopes botanists had documented a distinctive vegetation that might, as with Fiji's Des Voeux Peak, provide a refuge for such a bat. But Panié was far harder to reach than Des Voeux Peak. There was no road leading to the summit—only a steep track beginning at sea level. Arriving in the late morning, we engaged two young men from a local village as guides and set off through the tall, rank grass that clothed the mountain's lower slopes. The burrowing grass seeds, which worked their way into our clothing and boots, made it an unpleasant trek. And to add to our woes, the weather was unbearably humid.

With all of us carrying heavy packs, it would take until sunset to reach our destination—a resting hut a hundred metres or so below the summit. But after a few hours we crossed into the coolness of the forest, and the walk became enjoyable. By mid-afternoon we'd reached an elevation of 1000 metres, and had crossed an invisible line into a region where persistent mist, cloud and rain sit against the mountain. Alex had struck up a friendship with one of our guides, the son of a local chief, and who was also called Alex. As we rested on the mountain in an ancient glade, she murmured something in French. *'Oui, oui,'* answered the other Alex, and he continued in English, 'Yes. Yes, the spirits are here.'

The constant moisture had created an enchanted land filled with trees I didn't recognise. Some, I guessed, were akin to Australia's grevilleas, but they had flowers and leaves so huge that they appeared monstrous. Others were impossible for me to classify: they had waxen buds shaped like miniature spacecraft or new growth of the most fantastic shapes and colours. They were quite literally like nothing else on Earth: members of plant families that are unique to New Caledonia's mountains. Plant families are mostly ancient, and these plants had somehow survived on their island ever since it split from Australia. Since then a vegetable kingdom of such uniqueness had sprung up, making one dream of other worlds.

Towards the end of our climb we came to an abrupt low cliff, below which sprouted an abundance of metre-long, strap-like leaves with the most fantastic scarlet flowers waving at their tops. I recognised them as belonging to the lily family, and using our guidebook I soon identified them as *Xeronema moorei*, a rare plant found only in a few patches on New Caledonia's peaks. I was incredulous that this extraordinary plant, whose flower stalks more closely resemble vastly oversized scarlet toothbrushes than any lily I knew, seemed to have no common name. Surely the Kanaks must have stories about the blood-red blooms, I thought, but the young men we were with knew nothing of such things. It struck me as a sad waste that beauty like this has blossomed every year for millions of years while only rarely have human eyes drunk it in.

The hut that was to be our base had been built as a refuge against the weather by Europeans who knew nothing of the local conditions on the mountain. It was not insulated and lacked a place to make a fire. I've never seen a shelter so unsuited to requirements. As the gloom deepened and the warmth of the day gave way to a cold drizzle, supper beckoned. But we had nets to set before dark and paths to reconnoitre for spotlighting later on. The rain was falling steadily by the time we returned to the hut to heat a can of stew on our Trangia. We were exhausted, but the rough-hewn boards of the hut floor did not tempt us to neglect our duty. Instead, we walked back down to the 1000-metre mark again, spotlighting and checking the nets we'd set.

I wasn't in the best of moods as I slogged off through the steadily increasing rain, but as soon as I entered the forest both mood and exhaustion lifted. That night I saw things that I never imagined existed—I was in a *Through the Looking-Glass* world where life, shaped by ninety million years of splendid isolation, had created such hunters and hunted, plants and parasites they all might as well have originated in a distant galaxy.

Each leaf, each twig beside the path, was festooned with lichen and mosses big and small, their surfaces glistening in the torchlight with new-fallen droplets of rain. There were no possums, monkeys or indeed any terrestrial mammals in this world. Instead the leaves were grazed by slugs—huge, square, brightly patterned slugs, the largest ten centimetres from head to tail. They seemed to exist in every colour of the rainbow: some bright yellow with red lines, others grey with black lines, tan with yellow lines, or white with black.

The psychedelic molluscs moved through the silence of the night with all the solemn majesty of the moon, leaving in their wake silvery traces of their travels. Munching silently on the rotting vegetation that was their sustenance, they progressed so smoothly that their only movement seemed to come from the breathing-hole on their backs, which opened and closed in hypnotic slow motion.

Checking each mist-net and flowering tree, I saw nothing that might signal the presence of a bat. The forest was all but empty of large creatures, it seemed. But then the lion of this land in the clouds appeared. On a large leaf by the side of the path was a huntsman spider the size of my hand, its body glowing with an uncanny yellow phosphorescence. It sat imperturbable on its leaf—until I almost touched it. Puzzled by its indifference, I gently probed it with a stick, only to find that it had been transformed by an insidious, ramifying invader. The creature was not only dead, but glued to the leaf it sat upon. I levered it off to examine it more closely: its entire abdomen had been taken over by a fungus that glowed with the eerie phosphorescence that had first attracted me.

The fungal parasite must have somehow controlled the spider's mind, bidding it to climb on to the great leaf above the forest floor—an ideal perch from which to spread fungal spores. It had bade the spider sit still, exposed to predators, while it spread its hyphae from abdomen to leaf, forming a holdfast the spider would never break. And then it had taken the creature completely, its invisible hyphae reaching into each limb and section, until all that remained of the spider was an empty shell.

Presumably, in time the dead huntsman would sprout tiny

mushrooms, whose spores would blow free to infect yet other spiders. Just how a fungus can control a mind we cannot know. But I can tell you that high on forgotten mountains among distant islands, such things do occur.

That night sleep came late, and the dawn all too soon. After once again checking our nets, I summoned the energy to climb the hundred metres or so to the very summit of Mont Panié. The peak was in fact a small plateau, and in the scrubby habitat there grew a few dozen splendid pine trees, the largest thirty metres tall. They were *Araucaria schmidii*, an ancient type of pine belonging to a genus that last flourished during the age of dinosaurs. At that time such pines grew throughout the world, but today they exist only in Australia, South America and the islands of the south-west Pacific. *Araucaria schmidii* is arguably the rarest of all, being unique to the summit of Panié. Around me was its entire world population—just a few dozen plants.

After two nights atop the mountain, during which time it hardly stopped raining, we descended to the coast, checking our mist-nets as we went. The only mammal we found in any of them was a single specimen of the New Caledonian flying fox.

Evaluating such a result is difficult. Should we call it a failure? Certainly, based on such a brief survey, we can't rule out the possibility that a monkey-faced bat inhabits New Caledonia's highest peak. I think of the achievement as akin to leaving a few marks in the sand—marks that might one day guide a new generation of adventurers to hidden treasure.

Yet how to evaluate our work in the islands as a whole? To some, our adventures might seem to be nothing more than a romantic frolic. After all, why should anyone care about an obscure creature found only on a distant island? Would the world lose anything with its extinction? It can hardly be argued that island life is crucial to our own survival, for its impact on the maintenance of the Earth system must be immeasurably small. But there is more to life than mere survival. Who, after all, would not wish to see a living dodo? And how much richer would the economy of Mauritius be if that island still had, strutting around its forest, the astonishing and ridiculous *dodaars*, or knot-arses, as the seventeenth-century Dutch explorers lucky enough to see these gigantic, flightless pigeons in life called them?

The economic benefits of eco-tourism may be one reason for preserving island biodiversity, but for scientists there is a very different reward. Island species are of exceptional interest to anyone wishing to know how the evolutionary process works. That should include all of us, for evolution by natural selection is the force that shaped us and all the living world. If we hope to know ourselves we would do well to grasp its workings, and nowhere are they as intriguing or instructive as on islands.

During our sojourns on islands to the south and east of New Guinea we had investigated the influence of distance, island size, island age and isolation on the mammals whose ancestors had managed to reach these miniature worlds. We had discovered ten mammal species which were previously unknown to science. And, sadly, we had learned that some species documented by previous adventurers had almost certainly become extinct, among them the emperor rat of Guadalcanal. We also discovered that some, such as the giant rat of Malaita, had perished before they could even be documented and named by a scientist.

Importantly, as a result of our work, there was sufficient materi-al now to write the very first complete account of the mammals of the southwest Pacific region, making it easier for those who follow to target and undertake their own research work. We hoped that in time such work would add the evidence required to effectively conserve all of the region's mammals. Indeed our efforts had already laid the basis for some conservation initiatives. The International Union for the Conservation of Nature's rankings of species, which help determine which ones should be a priority for conservation action, relied on our accounts. We also wrote reports for the governments of the Pacific Island nations, documenting the richness of their biological heritage and the need for conservation. Unfortunately, such conservation is not a high priority for many of them.

Perhaps the greatest treasure of the Scott expeditions was the wealth of experience gathered by those who worked on the surveys. Wherever we went in the islands we encountered neglected histories. The impacts of nineteenth-century missionaries could be seen almost everywhere, in the form of the 'Mother Hubbard' dresses favoured by the local women and in the tiny churches tended by native pastors that dotted the islands. And it's impossible to ignore the fact that this was a setting for the greatest human conflict ever fought. The World War II battlegrounds of the Pacific campaign remain littered with airstrips, some almost as long as London's Heathrow, whole fleets of sunken warships and aircraft, armies of coral-crusted jeeps, trucks and tanks—indeed military *remanié* of all types. And, perhaps most memorably, we all experienced the kindness of strangers.

More than biodiversity has been lost to the world in the two short decades since our work was completed. In the aftermath of decolonisation, entire societies have been irrevocably altered by civil war and economic turmoil. We were lucky enough to see the Solomon Islands when Honiara was a thriving and harmonious town. We drank kava with chiefs in Fiji before they were caught up in military coups, and we camped on remote tropical beaches before they were overtaken by resort hotels. Perhaps we caught the islands in the last moments of a golden age before the twenty-first century intruded. At least that's what it feels like now.

Having completed our surveys of the southwest Pacific Islands, our work was far from over. Lying to the northwest and west of New Guinea is another great scatter of islands, on which evolution had led to the development of yet more unique biological realms. If we could extend our surveys into this area, a definitive history of the mammals of all of the Australasian islands could be written. This region lies within the great island archipelago of Indonesia, and it was there that, beginning in 1990, the Scott expeditions would concentrate. The wonders we were to discover in these western isles would eclipse even those of the southwest Pacific. But that is another story, the telling of which must await another book.

Afterword

Some readers may wonder why it was necessary for us to collect so many specimens for museums. After all, it necessitated the deaths of many individual animals, which may seem contradictory to our ultimate goal of the conservation of endangered species. It's important to understand that throughout the area in which we worked wild animals are an important food source. Where possible we took our samples (which usually included the skin, skull and liver) from individual animals that had been caught by local people for food. Where this was not possible, we collected samples sufficient to identify species using our own traps, nets and guns. Our techniques involved catching animals alive where possible, holding them in canvas bags until they could be examined and either released or humanely killed and sampled. Small animals needed as samples and not destined to be eaten by local people were by preference euthanased with a drop of Nembutal (which stops the heart) on the tongue, or an injection.

Why were so many samples required? The science of taxonomy, the classification of species, enables scientists to identify endangered species. Without it no conservation work can be done. Imagine being a biologist who realises that a population on an island is threatened with extinction. You want to help, but the first thing you need to do is to identify the species. This means that its fur and teeth need to be compared with the type specimens of similar species. Type specimens are the first individuals of a species to be named. They are like the flag-bearers for a species, and they are held in museum collections around the world. Having identified your species, you need to determine whether it is present on other islands, or only on the island in question. This requires adequate samples from all islands within the potential distribution of the species. Multiple samples are often required, for any island can be home to similar-looking species, which nonetheless are genetically and ecologically distinct.

During our surveys, we were often collecting the very first samples of a species from the island we were visiting. It was pioneering work in the vein of the nineteenth-century explorers. Anyone wishing to build on our work today could set about things very differently. Advances in DNA technology mean that identifications can now be made by sampling just a few hairs. Yet they would still rely heavily on comparisons with the extensive samples of skins, skulls and tissues that we collected over two decades ago.

For all its excitement and the value of the research, I don't think I could do the work today that I did back then. As I've got older I've found it harder and harder to kill animals, however good the cause.

It's hard to believe that twenty-five years have passed since I first voyaged to the islands. Back then I was a very different person— naive, filled with a restless youthful energy, and dangerously overconfident. I was also more careless, and I realise now with regret that my notebooks from the period reflect merely the bare bones of what happened—mainly where I went and what creatures we found. This story deals with far more than that, so is necessarily drawn heavily from memory, sometimes aided by photos taken at the time. To the best of my abilities it reflects my experiences, though of course others may recall things differently.

Among the Islands is not a straightforward recapitulation of what happened all those years ago. In order to maintain the narrative I've occasionally merged events that occurred over several expeditions. And because I've chosen a geographic arrangement of materials, the timing of various expeditions is sometimes left unstated, as a strict chronology would only confuse the reader. In some instances I've had to rely upon the accounts of others, who related their experiences to me soon after the events occurred. Where this is the case I make it clear in the text.

A small army of people needs to be thanked for their contributions to our efforts. The work could not have taken place without the generous support of the governments of Papua New Guinea, the Solomon Islands, Fiji and New Caledonia, particularly the ministries and departments involved in wildlife conservation. At the Australian Museum, I benefited from the wholehearted support of everyone, from the director Des Griffin down to the cleaners and guards, who worked so hard behind the scenes to make my job easier. All have my deepest thanks, as do the villagers of the regions I visited. Without their goodwill and assistance, nothing could have been achieved.

My greatest debt of gratitude lies with those who made up the Scott expedition team. We were never certain whether our funding would be renewed from year to year, and I appreciate the enormous patience and generosity of spirit required to go on in the face of the possibility that there'd be no money for salaries next season. Furthermore, all endured significant trials and risks in carrying out their work. Without exception they performed magnificently. The core team included: Ian Aujare, Tish Ennis, Dr Diana Fisher, Pavel German, Dr Sandra Ingleby, Tanya Leary, Peter Manueli, Dr Harry Parnaby, Lester Seri, Dr Alexandra Szalay and Dr Elizabeth Tasker. To them all, many thanks. What adventures we had!

References

1   Malinowski, B. *The Sexual Life of Savages in North-Western Melanesia*, George Routledge and Sons, London, 1929.

2   Meek, A. S. *A Naturalist in Cannibal Land*, T. Fischer Unwin, London, 1913, p. 76.

3   Ibid, p. 78.

4   Damon, F. *From Muyuw to the Trobriands*, University of Arizona Press, Tucson, 1990, p. 55.

5   Meek, A. S. *A Naturalist in Cannibal Land*, T. Fischer Unwin, London, 1913.

6   Brass, L. J. et al. 'Results of the Archbold Expeditions,' No. 75, 1956. 'Summary of the Fourth Archbold Expedition to New Guinea,' in *Bulletin of the American Museum of Natural History*, No. 111(2), 1953, p. 144.

7   Hempenstall, P. J. *Pacific Islands under German Rule*, ANU Press, Canberra, 1978, p. 151.

8   Troughton, E. Le G. and Livingstone, A. A. 'Last Days at Santa Cruz', *The Australian Zoologist*, 111(4), 1927, pp. 114–23.

9   Woodford, C. M. *A Naturalist Among the Head Hunters, Being an Account of Three Visits to the Solomon Islands in the Years 1886–1888*, George Phillip and Sons, London, 1890.

10 Hill, J. E. *A Memoir and Bibliography of Michael Rogers Oldfield Thomas, FRS*, British Museum of Natural History, Historical Series, 18(1), London, 1990, pp. 25–113.

11   Andersen, K. 'Diagnoses of new bats of the families Rhinolophidae and Megadermatidae', *Annals and Magazine of Natural History* 9(2), 1918.

12   Flannery, T. 'Stuffed & Pickled', *Australian Natural History*, 22(10), 1988.

13   Keesing, R. and Corris, P. *Lightning Meets the West Wind. The Malaita Massacre*, Oxford University Press, Melbourne and Oxford, 1980.

14   Ibid.

15   Ibid, p. 135.

16   Ibid, pp. 135–6.

17   Ibid, p. 188.

18   Ibid, p. 203.

19   Troughton, E. le G. 'The Mammalian Fauna of Bougainville Island, Solomons Group', *Records of the Australian Museum* XIX (5), 1936, p. 341.

20   Fisher, D. 'An Ecological Study of a New Species of Monkey-Faced Bat from the Islands of New Georgia and Vangunu, the Solomon Islands', Research report to the Mammal Department, Australian Museum, 1992.

21   Endicott, W. 'A Cannibal Feast at the Feejee Islands', *Danvers Courier*, 16 August, 1845; reprinted in: *Wrecked Among Cannibals in the Feejees*, Marine Research Society, Salem, Massachusetts, 1923.

22   Hill, A. V. S. and Serjeantson, S. W. (eds.) *The Colonisation of the Pacific. A Genetic Trail*, Clarendon Press, Oxford, 1989.

23   Ryan, P. *Fiji's Natural Heritage*, Exisle Publishing, NZ, 2000.

24   Williams T. and Calvert, J. *Fiji and the Fijians*, Fiji Museum, Suva, 1858, 1985 (reprint), p. 239.

25   Ibid.

26   Ibid.

27   Ibid, 152.

28   Ibid, 153.

29   Ibid.

30   Ibid, 155.

Index